About this book

This book is designed to provide you with the best preparation possible for your Edexcel S1 unit examination:

- The LiveText CD-ROM in the back of the book contains even more resources to support you through the unit.
- A matching S1 revision guide is also available.

Finding your way around the book

Brief chapter overview and 'links' to underline the importance of mathematics: to the real world, to your study of further units and to your career

Every few chapters, a review exercise helps you consolidate your learning

Detailed contents list shows which parts of the S1 specification are covered in each section

Each section begins with a statement of what is covered in the section

Concise learning points

Step-by-step worked examples

Past examination questions are marked 'E'

Each section ends with an exercise – the questions are carefully graded so they increase in difficulty and gradually bring you up to standard

Each chapter has a different colour scheme, to help you find the right chapter quickly

Each chapter ends with a mixed exercise and a summary of key points.

At the end of the book there is an examination-style paper.

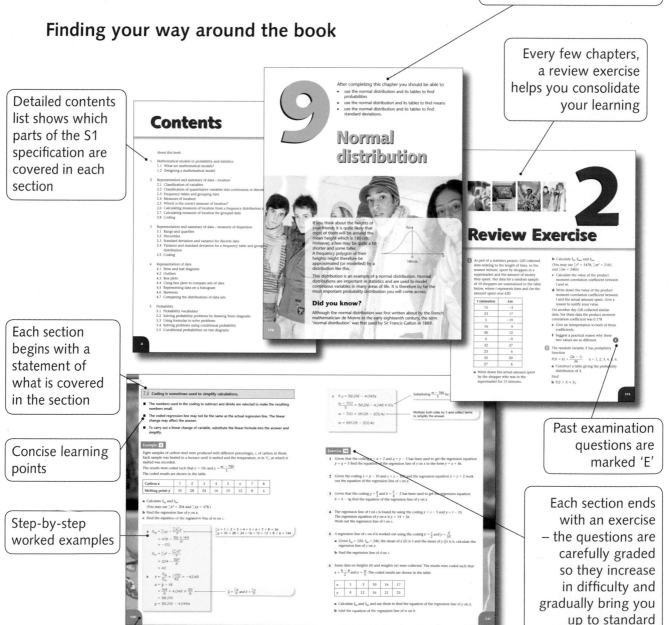

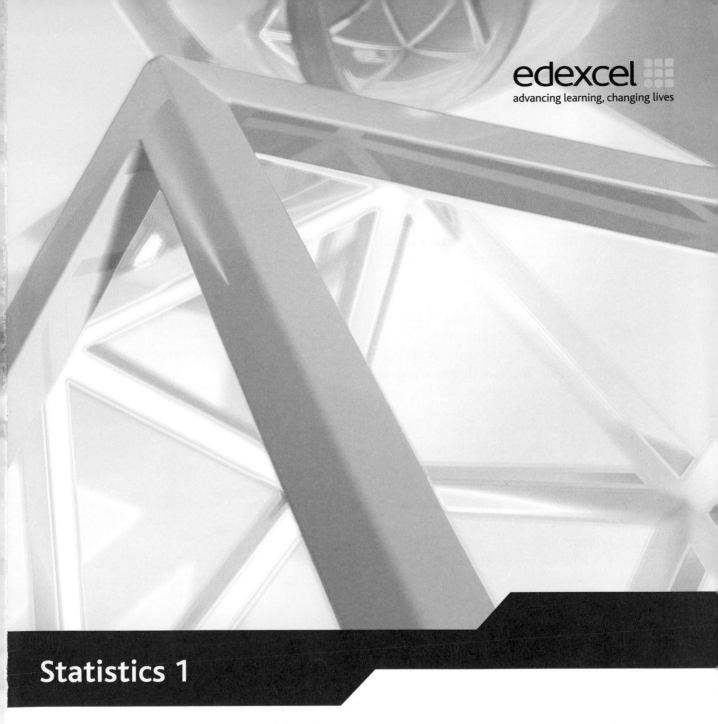

edexcel :::
advancing learning, changing lives

Statistics 1

Edexcel AS and A Level Modular Mathematics

Greg Attwood
Alan Clegg
Gill Dyer
Jane Dyer

Contents

LiveText software

The LiveText software gives you additional resources: Solutionbank and Exam café. Simply turn the pages of the electronic book to the page you need, and explore!

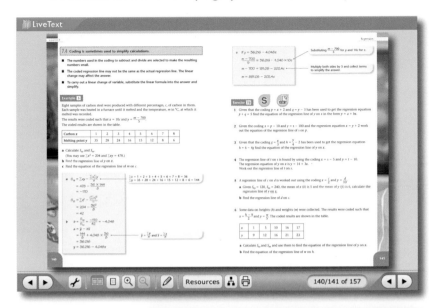

Unique Exam café feature:

- Relax and prepare – revision planner; hints and tips; common mistakes
- Refresh your memory – revision checklist; language of the examination; glossary
- Get the result! – fully worked examination-style paper

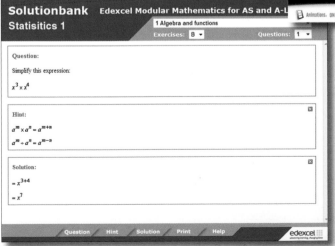

Solutionbank

- Hints and solutions to every question in the textbook
- Solutions and commentary for all review exercises and the practice examination paper

Pearson Education Limited, a company incorporated in England and Wales, having its registered office at 80 Strand, London, WC2R 0RL. Registered company number: 872828

Text © Greg Attwood, Gill Dyer, Jane Dyer, Alan Clegg 2008

15
17

British Library Cataloguing in Publication Data is available from the British Library on request.

ISBN 978 0 435519 124

Edited by Susan Gardner
Typeset by Tech-Set Ltd
Illustrated by Tech-Set Ltd
Cover design by Christopher Howson
Picture research by Chrissie Martin
Cover photo/illustration © Science Photo Library/Laguna Design
Printed in China (CTPS/17)

Acknowledgements

The author and publisher would like to thank the following individuals and organisations for permission to reproduce photographs:

Shutterstock / Khafizov Ivan Harisovich p1; Alamy Images / Vario Images GmbH & Co. KG p6; Shutterstock / iofoto p30; Alamy Images / Matthew Ashton p52; Alamy Images / Stock Connection Blue p77; Shutterstock / Ame Trautmann p114; Alamy Images / Niall McDiarmid p134; Getty Images / Photonica p150; Pearson Education Ltd / Studio 8, Clark Wiseman p176

Every effort has been made to contact copyright holders of material reproduced in this book. Any omissions will be rectified in subsequent printings if notice is given to the publishers.

Disclaimer

This Edexcel publication offers high-quality support for the delivery of Edexcel qualifications.

Edexcel endorsement does not mean that this material is essential to achieve any Edexcel qualifications, nor does it mean that this is the only suitable material available to support any Edexcel qualification. No endorsed material will be used verbatim in setting any Edexcel examination/assessment and any resource lists produced by Edexcel include this and other appropriate texts.

Copies of official specifications for all Edexcel qualifications may be found on the Edexcel website – www.Edexcel.com.

After completing this chapter you should be able to

- understand the process of mathematical modelling
- know the stages of the modelling process
- be able to discuss whether or not assumptions are reasonable.

Mathematics is not just about formulae and ideas. It has applications in most areas of modern day life. From engineering to medicine, or conservation to finance, mathematical modelling makes significant contributions to the world around us.

Mathematical models in probability and statistics

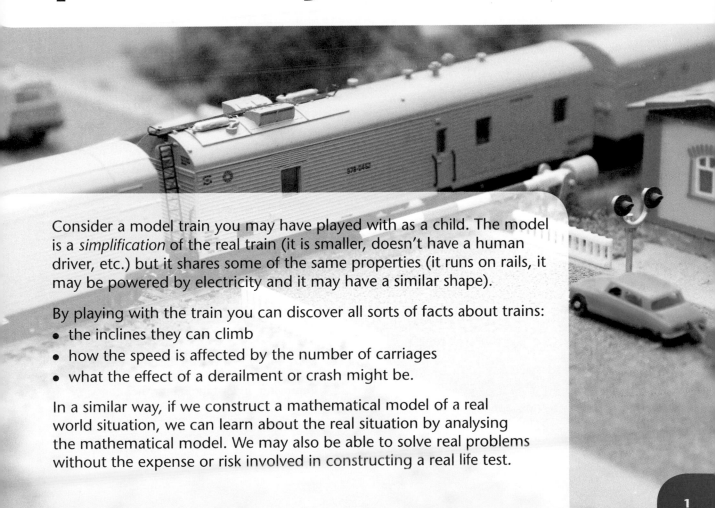

Consider a model train you may have played with as a child. The model is a *simplification* of the real train (it is smaller, doesn't have a human driver, etc.) but it shares some of the same properties (it runs on rails, it may be powered by electricity and it may have a similar shape).

By playing with the train you can discover all sorts of facts about trains:
- the inclines they can climb
- how the speed is affected by the number of carriages
- what the effect of a derailment or crash might be.

In a similar way, if we construct a mathematical model of a real world situation, we can learn about the real situation by analysing the mathematical model. We may also be able to solve real problems without the expense or risk involved in constructing a real life test.

1.1 Learn what mathematical models are, why they are useful and why they should sometimes be treated with caution.

A **mathematical model** is a **simplification** of a real world situation. It can be used to make **predictions** about a real world problem. By analysing the model an **improved understanding** of the situation may be obtained.

The model will aim to incorporate some but not all the features of the real situation. Certain assumptions have to be made, and these may mean the model does not share all the features of the real situation.

Mathematical models are useful because:

- they are **quick** and **easy** to produce
- they can **simplify** a more complex situation
- they can help us **improve our understanding** of the real world as certain **variables can readily be changed**
- they enable **predictions** to be made
- they can help **provide control** – as in aircraft scheduling.

Mathematical models should sometimes be treated with caution because:

- the model is a **simplification** of the real problem and **does not include all aspects** of the problem
- the model may **only work in certain situations**.

In chapter 7 you will learn about the dangers of extrapolation, which is using a model to provide predictions for values outside the range for which it is valid.

You will meet some mathematical models in the later chapters of this book and you will find other questions there which may ask you to comment on the suitability of a particular mathematical model in a practical situation.

Example **1**

Give two reasons for using mathematical models.

- Mathematical models are cheaper and easier to use than the real situation.
- Mathematical models can help improve our understanding of a real world problem.

Make sure you give two, distinct reasons: ease of use or cost for one reason and improving understanding for the other.

1.2 You can design a mathematical model.

You can split the process of designing a mathematical model into seven stages, represented by the following diagram.

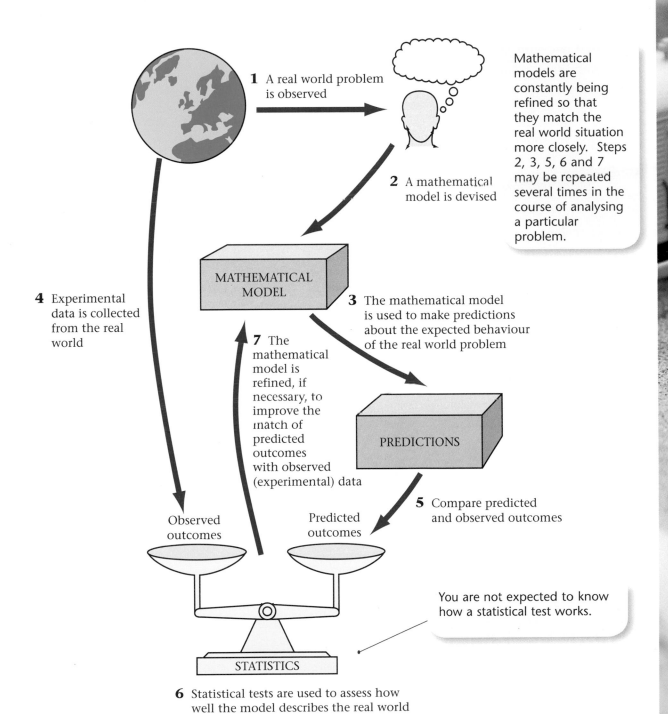

1 A real world problem is observed

2 A mathematical model is devised

Mathematical models are constantly being refined so that they match the real world situation more closely. Steps 2, 3, 5, 6 and 7 may be repeated several times in the course of analysing a particular problem.

MATHEMATICAL MODEL

4 Experimental data is collected from the real world

3 The mathematical model is used to make predictions about the expected behaviour of the real world problem

7 The mathematical model is refined, if necessary, to improve the match of predicted outcomes with observed (experimental) data

PREDICTIONS

Observed outcomes

Predicted outcomes

5 Compare predicted and observed outcomes

You are not expected to know how a statistical test works.

STATISTICS

6 Statistical tests are used to assess how well the model describes the real world

Statistics play an important role in this process. In chapters 2, 3 and 4 you will meet ways of processing experimental data and expected outcomes. In chapters 7, 8 and 9 you will be introduced to some simple mathematical models that incorporate the ideas of randomness or probability.

Example 2

A biologist notices that a population of birds fluctuates from year to year. Outline the stages that are needed to create a mathematical model to model this fluctuation.

> 1 Some assumptions are made, e.g. that birth rate and death rate should be included but food supply and temperature changes should not.
>
> 2 Devise a mathematical model – this may involve areas of pure mathematics as well as probability.
>
> 3 Use the model to predict the population over a period of several years.
>
> 4 Collect fresh data to match the conditions of the predicted values, or use historical data from preceding years.
>
> 5 Use some of the techniques you will meet in chapters 2, 3, 4 and 6 to compare the predicted values with the experimental data.
>
> 6 A statistical test provides an objective means of deciding if the differences between the model's predictions and the experimental data are within acceptable limits. Some examples of these will be seen in S2–S4.

If the predicted values did not match the experimental data well, the model can be refined. This involves repeating steps 2–6. For example, in this case the biologist may decide that the food supply should now be included in the model. New predictions can then be made and compared with experimental data.

Example 3

Explain briefly the role of statistical tests in the process of mathematical modelling.

> Statistical tests are used to assess how well a mathematical model matches a real world situation.

Exercise 1A

1 Give one advantage and one disadvantage of using a mathematical model.

2 Describe briefly the process of refining a mathematical model.

Summary of key points

1   A **mathematical model** is a **simplification** of a real world situation.

2   Some **advantages** of mathematical models are:
    - they are **quick** and **easy** to produce
    - they can **simplify** a more complex situation
    - they can help us **improve our understanding** of the real world as certain **variables can readily be changed**.
    - they enable **predictions** to be made about the future
    - they can help **provide control** – as in aircraft scheduling.

3   Some **disadvantages** of mathematical models are:
    - they only give a **partial** description of the real situation.
    - they only work for a **restricted** range of values.

After completing this chapter you should be able to

- recognise different types of data
- find the mean, mode and median for
  - discrete data presented as a list
  - discrete data presented in a table
  - continuous data presented in a grouped frequency table
- use coding to make calculations of measures of location simpler.

You will use these measures to compare distributions and decide which is most suitable for particular applications

Representation and summary of data – location

A farmer wishes to buy a laying hen. He asks the seller what the average number of eggs laid per week is. The seller has three measures he can use. He can quote the figure of eight eggs per week which is the median, 8.2 eggs per week which is the mean or 10 eggs per week which is the mode. Which measure should the seller quote?

2.1 The classification of variables depends on whether their observations can be written down as a number.

In **statistics** you collect **observations** or **measurements** of some **variable**. Such observations are known as **data.**

■ Variables associated with numerical observations are called **quantitative variables**.

■ Variables associated with non-numerical observations are called **qualitative variables**.

Example 1

For each of the variables in the table, state if their observations are numerical or not.

	Variable	Observations
1	Collar size	14, $14\frac{1}{2}$, 15, $15\frac{1}{2}$, 16
2	Height	177.8 cm, 160 cm, 180.4cm
3	Hair colour	Blonde, Red, Brunette

1 The observations of collar size are numerical.

You can give a number to collar size. Collar size is called a **quantitative** variable.

2 The observations of height are numerical.

You can give a number to height. Height is a **quantitative** variable.

3 The observations of hair colour are not numerical.

You can't give a number to hair colour. Hair colour is called a **qualitative** variable.

In S1 we shall be dealing mainly with quantitative variables.

2.2 The classification of quantitative variables depends on whether their observations are measured on a continuous or a discrete scale.

■ A variable that can take any value in a given range is a **continuous variable**.

■ A variable that can take only specific values in a given range is a **discrete variable**.

Example **2**

State whether or not each of the following variables is continuous or discrete.

a Time.　　　　　　**b** Length.　　　　　　**c** Number of 10p coins in a bag.

d Weight.　　　　　　**e** Number of girls in a family.

a	Time is continuous.
b	Length is continuous.
c	Number of 10p coins is discrete.
d	Weight is continuous.
e	Number of girls in a family is discrete.

Time can take any value. e.g. 2, 2.10, 2.01 etc.

You can't have 5.62 coins.

You can't have 2.65 girls in a family.

2.3 **Large amounts of discrete data can be written as a frequency table or as grouped data.**

Example **3**

Rebecca records the shoe size, x, of the female students in her year. The results are as follows.

x	Number of students, f
35	3
36	17
37	29
38	34
39	12

The number of anything is called its frequency. Here f stands for frequency.

A frequency table is a quick way of writing a long list of numbers. It tells us that three students had a shoe size of 35 and 17 students had a shoe size of 36, etc.

Find

a the number of female students who take shoe size 37,

b the shoe size taken by the smallest number of female students,

c the shoe size taken by the greatest number of female students,

d the total number of female students in the year.

a	29 students take shoe size 37.
b	The shoe size taken by the smallest number of students is 35.
c	The shoe size taken by the greatest number of students is 38.

The frequency of shoe size 37 is 29.

This has the lowest frequency.

This has the highest frequency.

d The total number of female students is 95.

> The sum of the frequencies is the total number of female students.
> 3 + 17 + 29 + 34 + 12 = 95.

It is sometimes helpful to add a column to the table showing the running total of the frequencies. This is called the **cumulative frequency**.

Example 4

Copy the table in Example 3 and add a cumulative frequency column.

x	Number of students, f	Cumulative frequency
35	3	3
36	17	20
37	29	49
38	34	83
39	12	95

> 3 + 17 = 20
>
> 20 + 29 = 49

Grouped data

When data is presented as a grouped frequency table, the specific data values are lost. You need to know the following.

- **The groups are more commonly known as classes.**
- **You need to be able to find the class boundaries.**
- **You need to be able to find the mid-point of a class.**
- **You need to be able to find the class width.**

Example 5

The length, x mm, to the nearest mm, of the forewing of a random sample of male adult butterflies is measured and shown in the table below. Write down the class boundaries, mid-point and class width for the class 34–36.

Length of forewing (mm)	Number of butterflies, f
30–31	2
32–33	25
34–36	30
37–39	13

> Class boundaries 33.5 mm, 36.5 mm

The data has gaps. Therefore the class boundaries are halfway between 33 and 34, etc.

> Mid-point $= \frac{1}{2}(33.5 + 36.5) = 35$ mm

To find the mid-point add the class boundaries and divide the result by two.

> Class width $= 36.5 - 33.5 = 3$ mm

The class width is the difference between the class boundaries.

Example 6

The time, x seconds, taken by a random sample of females to run 400 m is measured and shown in the table below. Write down the class boundaries, mid-point and class width for the class 70–75.

Time to run 400m (s)	Number of females f
55–65	2
65–70	25
70–75	30
75–90	13

As you can see the classes appear to overlap. This is not the case as 65–70 is the short form of $65 \leqslant x < 70$.

> Class boundaries 70s, 75s
> Mid-point $= \frac{1}{2}(70 + 75) = 72.5$s
> Class width $= 75 - 70 = 5$s

The data has no gaps therefore the class boundaries are the numbers of the class.

Exercise 2A

1 State whether each of the following variables is qualitative or quantitative.

   a Height of a tree.

   b Colour of car.

   c Time waiting in a queue.

   d Shoe size.

   e Name of pupils in a class.

2 State whether each of the following quantitative variables is continuous or discrete.

    a Shoe size.

    b Length of leaf.

    c Number of people on a bus.

    d Weight of sugar.

    e Time required to run 100 m.

    f Lifetime in hours of torch batteries.

3 Explain why

    a 'Type of tree' is a qualitative variable.

    b 'The number of pupils in a class' is a discrete quantitative variable.

    c 'The weight of a collie dog' is a continuous quantitative variable.

4 The nurse at a health centre records the heights, h cm, to the nearest cm, of a group of boys in the same school age group. The frequency table shows the results.

    a Complete the table by putting in the cumulative frequency totals.

    b State the number of boys who are less than 168 cm tall.

    c Write down the height that is the most common.

h	Frequency (f)	Cumulative frequency
165	8	
166	7	
167	9	
168	14	
169	18	
170	16	

5 The distribution of the lifetimes of torch batteries is shown in the grouped frequency table below.

Lifetime (nearest 0.1 of an hour)	Frequency	Cumulative frequency
5.0–5.9	5	
6.0–6.9	8	
7.0–7.9	10	
8.0–8.9	22	
9.0–9.9	10	
10.0–10.9	2	

    a Complete the cumulative frequency column.

    b Write down the class boundaries for the second group.

    c Work out the mid-point of the fifth group.

6 The distribution of the weights of two-month-old piglets is shown in the grouped frequency table below.

Weight (kg)	Frequency	Cumulative frequency
1.2–1.3	8	
1.3–1.4	28	
1.4–1.5	32	
1.5–1.6	22	

a Write down the class boundaries for the third group.

b Work out the mid-point of the second group.

7 Write down which of the following statements are true.

a The weight of apples is discrete data.

b The number of apples on the trees in an orchard is discrete data.

c The amount of time it takes a train to make a journey is continuous data.

d David collected data on car colours by standing at the end of his road and writing down the car colours. Is the data he collected qualitative?

2.4 **Three measures of location can be used to describe the centre of a set of data – mode, median and mean.**

A set of data can be summarised by giving a single number to describe its centre. This number is called a **measure of location** and is often described as an **average.** There are three alternative measures of location that you can use, namely the **mode, the median and the mean.** There are slightly different methods for calculating these measures for discrete and for continuous variables. In this section we are going to consider discrete data.

The mode

■ **The mode is the value that occurs most often.**

Example **7**

Find the mode of each of the following sets of data.

a   3   4   6   2   8   8   5

b   13   26   22   30   32   48   29   27   26   32

c   6   8   4   2   9   3   7

a Mode = 8	8 occurs twice, the other numbers only occur once.
b Mode = 26 and 32	Both 26 and 32 occur twice. There are two modes for this data. This is called bimodal.
c There is no mode	All numbers have the same frequency.

The median

■ The median is the middle value when the data is put in order.

■ In general if there are n observations first divide n by 2. When $\frac{n}{2}$ is a whole number find the mid-point of the corresponding term and the term above. When $\frac{n}{2}$ is not a whole number round the number up and pick the corresponding term.

Example 8

Find the median of each of the following sets of data.

a   3   4   6   2   8   8   5

b   13   26   22   30   32   48   29   27   26   32

a Putting the numbers in order
    2 3 4 **5** 6 8 8
    Median = 5

Here you can see the middle value. There are 7 values and $\frac{7}{2}$ = 3.5 so rounding up gives 4. The median is therefore the fourth term when the list is ordered.

b Putting the numbers in order.
    13 22 26 26 **27 29** 30 32 32 48
    Median = $\frac{(27 + 29)}{2}$ = 28

There are 10 values and $\frac{10}{2}$ = 5. This is a whole number so we take the mid-point of the fifth and sixth terms.

The mean

■ The mean is the sum of all the observations divided by the total number of observations.

■ You can use the symbol Σ (the Greek letter s) instead of writing 'the sum of', and you can use $\bar{x}$ instead of writing 'the mean of the observations $x_1, x_2 \ldots x_n$'.

■ The mean is given by

$$\text{Mean} = \frac{\text{Sum of observations}}{\text{Number of observations}} = \frac{\Sigma x}{n}$$

$\bar{x}$ is the symbol for the mean of a sample. μ is the symbol used for the mean of a population.

Example 9

Find the mean of the following set of data.

$$3 \quad 4 \quad 6 \quad 2 \quad 8 \quad 7 \quad 5$$

$$\text{The mean} = \frac{3 + 4 + 6 + 2 + 8 + 7 + 5}{7}$$

$$= 5$$

Combining means

If set A, of size n_1, has mean $\bar{x}_1$ and set B, of size n_2, has a mean $\bar{x}_2$, then the mean of the combined set of A and B is

$$\bar{x} = \frac{n_1\bar{x}_1 + n_2\bar{x}_2}{n_1 + n_2}$$

Example 10

The mean of a sample of 25 observations is 6.4. The mean of a second sample of 30 observations is 7.2. Calculate the mean of all 55 observations.

The total sum of the 25 observations

$$= 25 \times 6.4$$

The total sum of the 30 observations

$$= 30 \times 7.2$$

$$\text{Mean} = \frac{25 \times 6.4 + 30 \times 7.2}{55}$$

$$= 6.84 \text{ (to 2 d.p.)}$$

Remember you need the total sum of the observations divided by the total number of observations. **You must not** just add the two means.

Example 11

A child at a junior school records the maximum temperature, in °C, for seven days at his school. The results are given below.

$$15.7 \quad 16.1 \quad 16.2 \quad 47.6 \quad 17.4 \quad 18.6 \quad 16.7$$

a Find the mean and median of these data.

The child's teacher realises that the figure 47.6 should be 17.6

b Write down what effect this will have on the median and mean.

a   Mean $= \dfrac{\sum x}{n}$

$= \dfrac{15.7 + 16.1 + 16.2 + 47.6 + 17.4 + 18.6 + 16.7}{7}$

$= 21.2$

15.7, 16.1, 16.2, **16.7**, 17.4, 18.6, 47.6

Median $= 16.7$

> Looking at the data given, the mean is higher than all but one piece of data. It is not a good measure of these data. The median is more representative.

b   The median will stay the same but the mean will decrease to 16.9.

Mean $= \dfrac{148.3 - 30}{7} = 16.9$

Exercise 2B

1   Meryl collected wild mushrooms every day for a week. When she got home each day she weighed them to the nearest 100 g. The weights are shown below.

500   700   400   300   900   700   700

  a Write down the mode for these data.

  b Calculate the mean for these data.

  c Find the median for these data.

On the next day, Meryl collects 650 g of wild mushrooms.

  d Write down the effect this will have on the mean, the mode and the median.

2   Joe collects six pieces of data x_1, x_2, x_3, x_4, x_5 and x_6. He works out that $\sum x$ is 256.2.

  a Calculate the mean for these data.

He collects another piece of data. It is 52.

  b Write down the effect this piece of data will have on the mean.

3   A small workshop records how long it takes, in minutes, for each of their workers to make a certain item. The times are shown in the table.

Worker	A	B	C	D	E	F	G	H	I	J
Time in minutes	7	12	10	8	6	8	5	26	11	9

  a Write down the mode for these data.

  b Calculate the mean for these data.

  c Find the median for these data.

  d The manager wants to give the workers an idea of the average time they took. Write down, with a reason, which of the answers to **a**, **b** and **c** he should use.

4 A farmer keeps a record of the weekly milk yield of his herd of seven cows over a period of six months. He finds that the mean yield is 24 litres. He buys another cow that he is told will produce 28 litres of milk a week. Work out the effect this will have on the mean milk yield of his cows.

5 A clothes retailer has two shops in the town of Field-gate. Shop A employs 15 people and shop B employs 22 people. The mean number of days of sickness in a year taken by the employees in shop A is 4.6 and the mean number of days of sickness taken by the employees in shop B is 6.5 days. Calculate the mean number of days of sickness taken per year by all 37 employees. Give your answer to one decimal place.

6 The rainfall in a certain seaside holiday resort was measured, in millimetres, every week for ten weeks. The hours of sunshine were also recorded. The data are shown in the table.

Rainfall (mm)	0	1	2	3	3	26	3	2	3	0
Sunshine (hours)	70	15	10	15	18	0	15	21	21	80

a Calculate the mean rainfall per week.

b Calculate the mean number of hours of sunshine per week.

c Write down the modal amount of rainfall and the modal amount of sunshine per week.

d Work out the median rainfall and the median amount of sunshine per week.

The council plans to produce a brochure and in it they wish to promote the resort as having lots of sunshine and little rain.

e Write down, with reasons, which of the mean, mode or median they should quote in their brochure as the average rainfall and hours of sunshine.

7 The mean marks for a statistics exam were worked out for three classes. Class 1 had 12 students with a mean mark of 78%. Class 2 had 16 students with a mean mark of 84%. Class 3 had 18 students with a mean mark of 54%. Work out the mean % mark to the nearest whole number for all 46 students.

2.5 **You should know which is the correct measure of location to use in each situation.**

- **Mode** This is used when data is qualitative, or quantitative with either a single mode or bimodal. It is not very informative if each value occurs only once. The mode was not asked for in example 11 as every number occurs once only.

- **Median** This is used for quantitative data. It is usually used when there are extreme values, as in example 11, as they do not affect it.

- **Mean** This is used for quantitative data and uses all the pieces of data. It therefore gives a true measure of the data. However, it is affected by extreme values.

The measure you use will depend on what you are trying to achieve. You will need to be able to decide on the best measure to use in particular situations.

Example 12

A company consists of seven workers paid at £10 per hour and their supervisor who is paid at £50 per hour.

a Find the mode, median and mean of all eight workers.

Write down, with a reason, which of the mean, mode and median you would use in the following situations.

b When asked the typical hourly rate of pay for the company.

c When trying to persuade a prospective employee to work for the company.

> **a** Mode = £10
>
> Median = £10
>
> Mean = $\dfrac{7 \times 10 + 50}{8}$ = £15
>
> **b** The mode or median, £10, as it gives the value most people get.
>
> **c** The mean, £15, as it is a higher value and more likely to persuade the prospective employee.

2.6 You can calculate the mean, mode and median for data presented in a frequency distribution table.

■ **The mean of a sample of data, that is summarised as a frequency distribution, is $\bar{x}$ where:**

$$\bar{x} = \frac{\sum fx}{\sum f}$$

Example 13

Rebecca records the shirt collar size, x, of the male students in her year. The results are as follows.

x	Number of students, f
15	3
15.5	17
16	29
16.5	34
17	12

A frequency table is a quick way of writing a long list of numbers. It tells us that three students had a collar size of 15 and seventeen students had a collar size of 15.5 etc.

Find for these data

a the mode   **b** the median   **c** the mean.

a mode = 16.5 •

> 16.5 is the collar size with the highest frequency.

To find the median and mean it is helpful to add a column for the **cumulative frequency**.

x	Number of students	Cumulative frequency
15	3	3
15.5	17	20
16	29	49
16.5	34	83
17	12	95

> The total number of observations is the total of the frequencies. i.e. the last cumulative frequency total, 95.
> Since $\frac{95}{2} = 47.5$, rounding up gives 48 so the median value is the 48th term. The 20th observation is 15.5 and the 49th term is 16 with every observation in between having a value of 16.

b median = 16 •

c mean

$= \dfrac{15 \times 3 + 15.5 \times 17 + 16 \times 29 + 16.5 \times 34 + 17 \times 12}{95}$ •

$= \dfrac{45 + 263.5 + 464 + 561 + 204}{95} = \dfrac{1537.5}{95} = 16.2$

> The total of the observations is found by multiplying each observation by the frequency with which it occurs.

Exercise 2C

1 The marks scored in a multiple choice statistics test by a class of students are:

5	9	6	9	10	6	8	5	5	7	9	7
8	6	10	10	7	9	6	9	7	7	7	8
6	9	7	8	6	7	8	7	9	8	5	7

a Draw a frequency distribution table for these data.

b Calculate the mean mark for these data.

c Write down the number of students who got a mark greater than the mean mark.

d Write down whether or not the mean mark is greater than the modal mark.

2 The table shows the number of eggs laid in 25 blackbirds' nests.

Number of eggs	0	1	2	3	4	5	6	7
Number of nests	0	0	0	1	3	9	8	4

Using your knowledge of measures of location decide what number of eggs you could expect a blackbird's nest to contain. Give reasons for your answer.

3 The table shows the frequency distribution for the number of petals in the flowers of a group of celandines.

a Work out how many celandines were in the group.

b Write down the modal number of petals.

c Calculate the mean number of petals.

d Calculate the median number of petals.

e If you saw a celandine, write down how many petals you would expect it to have.

Number of petals	Frequency (f)
5	8
6	57
7	29
8	3
9	1

4 The frequency table shows the number of breakdowns, b, per month recorded by a road haulage firm over a certain period of time.

Breakdowns b	Frequency f	Cumulative frequency
0	8	8
1	11	19
2	12	31
3	3	34
4	1	35
5	1	36

a Write down the number of months for which the firm recorded the breakdowns.

b Write down the number of months in which there were two or fewer breakdowns.

c Write down the modal number of breakdowns.

d Find the median number of breakdowns.

e Calculate the mean number of breakdowns.

f In a brochure about how many loads reach their destination on time, the firm quotes one of the answers to **c**, **d** or **e** as the number of breakdowns per month for its vehicles. Write down which of the three answers the firm should quote in the brochure.

5 A company makes school blazers in eight sizes. The first four sizes cost £48. The next three sizes cost £60 and the largest size costs £76.80. Write down, with a reason which of the mean, mode, or median cost the company is likely to use in advertising its average price.

2.7 The mean, modal class and median can be calculated for grouped data. This section deals with grouped data. Original data may have been discrete or continuous, but all grouped data is treated as continuous data.

When data is presented as a grouped frequency table, specific data values are lost. This means that we work out **estimates** for the mean and median and the modal class rather than a specific value.

■ The mean of a sample of data that is summarised as a grouped frequency distribution is $\bar{x}$ where:

$$\bar{x} = \frac{\sum fx}{\sum f} \text{ and } x \text{ is the mid-point of the group}$$

> We do not need to do any rounding as we are dealing with continuous variables.

■ To find the median divide n by 2 and use interpolation (see page 21) to find the value of the corresponding term.

■ The modal class is the class with the highest frequency.

> Since we do not have specific data values we call the class with the highest frequency the modal class.

Example 14

The length x mm, to the nearest mm, of a random sample of pine cones is measured. The data is shown below.

Length of pine cone (mm)	Number of pine cones, f	Cumulative frequency
30–31	2	2
32–33	25	27
34–36	30	57
37–39	13	70

a Write down the modal class.
b Estimate the mean length of pine cone.
c Estimate the median length of pine cone.

a modal class = 34–36

b mean = $\dfrac{30.5 \times 2 + 32.5 \times 25 + 35 \times 30 + 38 \times 13}{70}$

= 34.54

> For grouped data the total of the observations is found by multiplying the mid-point of each interval by the frequency of that interval.

c The median, m, is the $\dfrac{70}{2} = 35$th value.

This lies in the class 34–36 but we do not know the exact value of the 35th term.

To find an estimate for the median we need to use **interpolation**. First set up the diagram below.

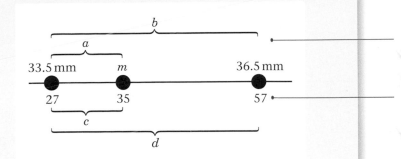

The end values on top are the class boundaries.

The end values on the bottom are the cumulative frequency for the previous class and this class.

$$\frac{m - 33.5}{36.5 - 33.5} = \frac{35 - 27}{57 - 27}$$

$$\frac{m - 33.5}{3} = \frac{8}{30}$$

The two fractions $\frac{a}{b}$ and $\frac{c}{d}$ are equivalent; therefore we can form the equation $\frac{a}{b} = \frac{c}{d}$ and solve to find the value of m.

$$m - 33.5 = \frac{8}{30} \times 3$$

$$m = 33.5 + 0.8$$

$$= 34.3 \qquad \text{median} = 34.3$$

Always check your answer is in the required range. i.e in this case between 33.5 and 36.5.

Example 15

The numbers of questions answered correctly by children taking a general knowledge test are shown in the following frequency distribution.

Number of correct answers	Frequency
0–5	4
6–10	15
11–15	5
16–20	2
21–60	0
61–70	1

Estimate

a the mean number of correct answers,

b the median number of correct answers.

c State whether the median or the mean is a better representation of the number of correct answers. Give a reason for your answer.

a Mean = $\dfrac{2.5 \times 4 + 8 \times 15 + 13 \times 5 + 18 \times 2 + 65.5 \times 1}{27}$

= 10.98

b The median is the $\dfrac{27}{2}$ = 13.5th value which lies in the interval 5.5 – 10.5

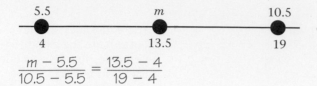

$$\frac{m - 5.5}{10.5 - 5.5} = \frac{13.5 - 4}{19 - 4}$$

$$\frac{m - 5.5}{5} = \frac{9.5}{15}$$

$$m - 5.5 = 3.166666$$

$$m = 8.666$$

median = 8.67

> In this case the number of correct marks is a discrete variable which in this situation has been treated as a continuous variable. While not ideal, it is the best you can do.

c The median is better. There is an extreme value and, while the median is not affected by extreme values, the mean is.

Exercise 2D

1 A hotel is worried about the reliability of its lift. It keeps a weekly record of the number of times it breaks down over a period of 26 weeks. The data collected are summarised in the table opposite.

a Estimate the mean number of breakdowns.

b Use interpolation to estimate the median number of breakdowns.

Number of breakdowns	Frequency of breakdowns (f)
0–1	18
2–3	7
4–5	1

c The hotel considers that an average of more than one breakdown per week is not acceptable. Judging from your answers to **b** and **c** write down with a reason whether or not you think the hotel should consider getting a new lift.

2 The weekly wages (to the nearest £) of the production line workers in a small factory is shown in the table.

a Write down the modal class.

b Calculate an estimate of the mean wage.

c Use interpolation to find an estimate for the median wage.

Weekly wage £	Number of workers, f,
175–225	4
226–300	8
301–350	18
351–400	28
401–500	7

3 The noise levels at 30 locations near an outdoor concert venue were measured to the nearest decibel. The data collected is shown in the grouped frequency table.

Noise (decibels)	65–69	70–74	75–79	80–84	85–89	90–94	95–99
Frequency (*f*)	1	4	6	6	8	4	1

   a Calculate an estimate of the mean noise level.

   b A noise level above 82 decibels was considered unacceptable. Estimate the number of locations that had unacceptable noise levels.

4 DIY store **A** considered that it was good at employing older workers. A rival store **B** disagreed and said that it was better. The two stores produced a frequency table of their workers' ages. The table is shown below

Age of workers (to the nearest year)	Frequency store A	Frequency store B
16–25	5	4
26–35	16	12
36–45	14	10
46–55	22	28
56–65	26	25
66–75	14	13

By comparing estimated means for each store decide which store employs more older workers.

5 The speeds of vehicles passing a checkpoint were measured over a period of one hour, to the nearest mph. The data collected is shown in the grouped frequency table.

Speed (mph)	21–30	31–40	41–50	51–60	61–65	66–70	71–75
No. of vehicles (*f*)	4	7	38	42	5	3	1

   a Write down the modal class.

   b Calculate the difference, to two decimal places, between the median and the mean estimated speeds.

   c The speed limit on the road is 60 mph. Work out an estimate for the percentage of cars that exceeded the speed limit.

2.8 **When data values are large, coding makes the numbers easier to work with.**

- There are many different ways of **coding**. You will normally be told the form that you should use. Coding is normally of the form

$$y = \frac{x - a}{b}$$

where a and b are to be chosen.

- To find the mean of the original data; find the **mean of the coded data**, equate this to the coding used and solve.

Example 16

a Find the mean of the following lengths, x mm.

110   120   130   140   150

b Use the following coding to find the mean of these data.

i $y = \frac{x}{10}$ **ii** $y = x - 100$ **iii** $y = \frac{x - 100}{10}$

a   mean $= 130$

To find coded data, divide each piece of data by 10 since the coding is $\frac{x}{10}$.

b   **i**   coded data 11, 12, 13, 14, 15,

mean of coded data $= 13$

mean of original data: $13 = \frac{x}{10}$

$x = 130$

$\dfrac{11 + 12 + 13 + 14 + 15}{5} = 13.$

Put the mean of the coded data equal to the coding and solve.

ii   coded data 10, 20, 30, 40, 50

mean of coded data $= 30$

mean of original data: $30 = x - 100$

$x = 130$

iii   coded data 1, 2, 3, 4, 5

mean of coded data $= 3$

mean of original data: $3 = \frac{x - 100}{10}$

$x = 130$

Example 17

Use coding to estimate the mean length of a telephone call, l, given the data in the table below.

Length of telephone call	Number of occasions
$0 < l \leqslant 5$	4
$5 < l \leqslant 10$	15
$10 < l \leqslant 15$	5
$15 < l \leqslant 20$	2
$20 < l \leqslant 60$	0
$60 < l \leqslant 70$	1

Use the coding $y = \dfrac{x - 7.5}{5}$
where x is the mid-point of each class

7.5 was chosen as it is the mid-point of the modal class. 5 was chosen as this is the smallest class width but any numbers would do. You will usually be given the coding in the question.

Length of telephone call	Number of occasions	Mid-point x	$y = \dfrac{x - 7.5}{5}$
$0 < l \leqslant 5$	4	2.5	−1
$5 < l \leqslant 10$	15	7.5	0
$10 < l \leqslant 15$	5	12.5	1
$15 < l \leqslant 20$	2	17.5	2
$20 < l \leqslant 60$	0	40	6.5
$60 < l \leqslant 70$	1	65	11.5

The table shows how to carry out calculations.

Mean of coded data

$$= \frac{-1 \times 4 + 0 \times 15 + 1 \times 5 + 2 \times 2 + 11.5 \times 1}{27}$$

$$= 0.61111111$$

$$0.611111 = \frac{x - 7.5}{5}$$

$$0.61111 \times 5 = x - 7.5$$

$$x = 10.56$$

Example 18

Data is coded using $y = \dfrac{x - 150}{50}$. The mean of the coded data is 36.2. Find the mean of the original data.

$36.2 = \dfrac{x - 150}{50}$

$36.2 \times 50 = x - 150$

$x = 1960$

Exercise 2E

1 Calculate the mean of the following data set (x) using the coding $y = \dfrac{x}{10}$.

  110   90   50   80   30   70   60

2 Find the mean of the following data set (x) using the coding $y = \dfrac{x - 3}{7}$.

  52   73   31   73   38   80   17   24

3 a Calculate the mean of 1, 2, 3, 4, 5 and 6.

Using your answer to **a**:

b Write down the mean of:

   i 2, 4, 6, 8, 10 and 12,

   ii 10, 20, 30, 40, 50 and 60,

   iii 12, 22, 32, 42, 52 and 62.

4 The coded mean price of televisions in a shop was worked out. Using the coding $y = \dfrac{x - 65}{200}$ the mean price was 1.5. Find the true mean price of the televisions.

5 The grouped frequency table shows the age (a years) at which a sample of 100 women had their first child.

Age of Women (a years)	Frequency (f)	Mid-point (x)	$y = \dfrac{x - 14}{2}$
11–21	11		
21–27	24		
27–31	27		
31–37	26		
37–43	12		

a Copy and complete the table

b Use the coding $y = \dfrac{x - 14}{2}$ to calculate an estimate of the mean age at which women have their first child.

Mixed exercise 2F

1 The following figures give the number of children injured on English roads each month for a certain period of seven months.

55   72   50   66   50   47   38

  a Write down the modal number of injuries.   **b** Find the median number of injuries.

  c Calculate the mean number of injuries.

2 The mean Science mark for one group of eight students is 65. The mean mark for a second group of 12 students is 72. Calculate the mean mark for the combined group of 20 students.

3 A computer operator transfers an hourly wage list from a paper copy to her computer. The data transferred is given below:

£5.50   £6.10   £7.80   £6.10   £9.20   £91.00   £11.30

  a Find the mean, mode and median of these data.

  The office manager looks at the figures and decides that something must be wrong.

  b Write down, with a reason, the mistake that has probably been made.

4 A piece of data was collected from each of nine people. The data was found to have a mean value of 35.5. One of the nine people had given a value of 42 instead of a value of 32.

  a Write down the effect this will have had on the mean value.

  The correct data value of 32 is substituted for the incorrect value of 42.

  b Calculate the new mean value.

5 On a particular day in the year 2007, the prices (x) of six shares were as follows:

807   967   727   167   207   767

  Use the coding $y = \dfrac{x - 7}{80}$ to work out the mean value of the shares.

6 The coded mean of employee's annual earnings (£x) for a store is 18. The code used was

  $y = \dfrac{x - 720}{1000}$. Work out the uncoded mean earnings.

7 Different teachers using different methods taught two groups of students. Both groups of students sat the same examination at the end of the course. The students' marks are shown in the grouped frequency table.

  a Work out an estimate of the mean mark for Group A and an estimate of the mean mark for Group B.

  b Write down whether or not the answer to **a** suggests that one method of teaching is better than the other. Give a reason for your answer.

Exam Mark	Frequency Group A	Frequency Group B
20–29	1	1
30–39	3	2
40–49	6	4
50–59	6	13
60–69	11	15
70–79	10	6
80–89	8	3

8 The table summarises the distances travelled by 150 students to college each day.

a Use interpolation to calculate the median distance for these data.

The mid-point is x and the corresponding frequency is f. Calculations give the following values

$\sum fx = 1056$

b Calculate an estimate of the mean distance for these data.

Distance (nearest km)	Number of students
0–2	14
3–5	24
6–8	70
9–11	32
12–14	8
15–17	2

9 Chloe enjoys playing computer games. Her parents think that she spends too much time playing games. Chloe decides to keep a record of the amount of time she plays computer games each day for 50 days. She draws up a grouped frequency table to show these data.

The mid-point of each class is represented by x and its corresponding frequency is f. (You may assume $\sum fx = 1275$)

a Calculate an estimate of the mean time Chloe spends on playing computer games each day.

Time to the nearest minute	Frequency
10–14	1
15–19	5
20–24	18
25–29	15
30–34	7
35–39	3
40–44	1

Chloe's parents thought that the mean was too high and suggested that she try to reduce the time spent on computer games. Chloe monitored the amount of time she spent on games for another 40 days and found that at the end of 90 days her overall mean time was 26 minutes.

b Find an estimate for the mean amount of time Chloe spent on games for the 40 days.

c Comment on the two mean values.

10 The lifetimes of 80 batteries, to the nearest hour, is shown in the table opposite.

a Write down the modal class for the lifetime of the batteries.

b Use interpolation to find the median lifetime of the batteries.

The mid-point of each class is represented by x and its corresponding frequency by f, giving $\sum fx = 1645$.

c Calculate an estimate of the mean lifetime of the batteries.

Lifetime (hours)	Number of batteries
6–10	2
11–15	10
16–20	18
21–25	45
26–30	5

Another batch of 12 batteries is found to have an estimated mean lifetime of 22.3 hours.

d Find the mean lifetime for all 92 batteries.

Summary of key points

1. A variable that can take any value in a given range is a **continuous** variable.

2. A variable that can take only specific values in a given range is a **discrete** variable.

3. A **grouped frequency distribution** consists of **classes** and their related **class frequencies**

   Classes   30–31   **32–33**   34–35
   For the class 32–33
   Lower class boundary   is   31.5
   Upper class boundary   is   33.5
   Class width   is $33.5 - 31.5 = 2$
   Class mid-point   is $\frac{1}{2}(31.5 + 33.5) = 32.5$

4. The **mode** or **modal class** is the value or class that occurs most often.

5. The **median** is the middle value when the data is put in order.

6. For **discrete** data divide n by 2. When $\frac{n}{2}$ is a whole number find the mid-point of the corresponding term and the term above. When $\frac{n}{2}$ is not a whole number, round the number up and pick the corresponding term.

7. For **continuous** grouped data divide n by 2 and use interpolation to find the value of the corresponding term.

8. The **mean** is given by

   $\bar{x} = \frac{\sum x}{n}$ or $\frac{\sum fx}{\sum f}$ (for grouped data x is the mid-point of the class).

9. If set A, of size n_1, has mean $\bar{x}_1$ and set B, of size n_2, has a mean $\bar{x}_2$ then the **mean of the combined** set of A and B is $\bar{x} = \frac{n_1\bar{x}_1 + n_2\bar{x}_2}{n_1 + n_2}$.

10. When the data values are large, you can use coding to make the numbers easier to work with. To find the mean of the original data, find the mean of the coded data, equate this to the coding used and solve.

After completing this chapter you should be able to

- find the **quartiles**, **range**, **interquartile range**, **variance** and **standard deviation** for: discrete data presented as a list, discrete data presented in a table, continuous data presented in a grouped frequency table
- use **coding** to make calculations of measures of dispersion simpler.

Representation and summary of data – measures of dispersion

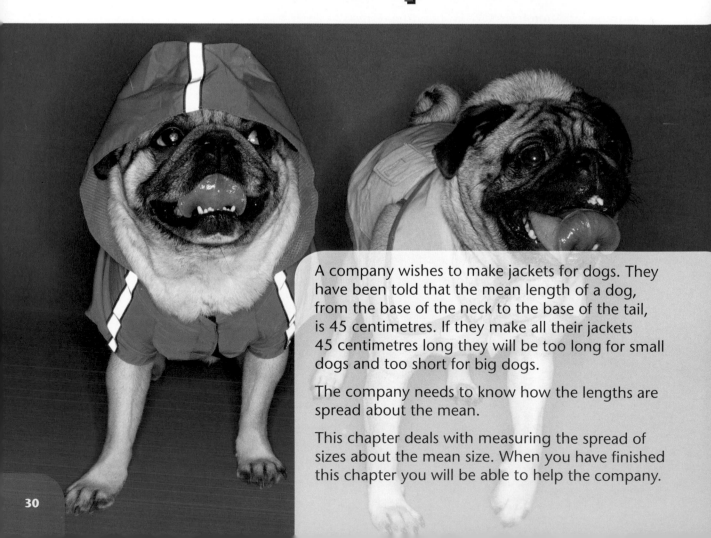

A company wishes to make jackets for dogs. They have been told that the mean length of a dog, from the base of the neck to the base of the tail, is 45 centimetres. If they make all their jackets 45 centimetres long they will be too long for small dogs and too short for big dogs.

The company needs to know how the lengths are spread about the mean.

This chapter deals with measuring the spread of sizes about the mean size. When you have finished this chapter you will be able to help the company.

3.1 **How to calculate the range, find quartiles and use them to work out the interquartile range.**

Range

The range of a set of data is very simple to find.

range = highest value − lowest value

Quartiles

Quartiles, Q_1, Q_2, Q_3, split the data into four parts.

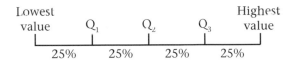

25% of the observations should have a value less than the lower quartile, Q_1, 50% of the observations should have a value below the median, Q_2 and 75% of the observations should have a value below the upper quartile, Q_3.

There are many different methods for calculating the quartiles and they may give slightly different answers. In this book we will use the following method where in general, there are n values and the numbers are in ascending order.

Method used for discrete data

This method is the same as we used for finding the median in Chapter 2.

- To calculate the **lower quartile**, Q_1, divide n by 4. When $\frac{n}{4}$ is a whole number find the mid-point of the corresponding term and the term above. When $\frac{n}{4}$ is not a whole number round the number up and pick the corresponding term.

- To calculate the **upper quartile**, Q_3, divide n by four and multiply by three. When $\frac{3n}{4}$ is a whole number find the mid-point of the corresponding term and the term above. When $\frac{3n}{4}$ is not a whole number round the number up and pick the corresponding term.

Method used for continuous data

- To calculate the **lower quartile**, Q_1, divide n by four and use interpolation to find the corresponding value.

- To calculate the **upper quartile**, Q_3, divide n by four and multiply by three and use interpolation to find the corresponding value.

Interquartile range

interquartile range = upper quartile − lower quartile

Example 1

Find the range and interquartile range of the following data.

7   9   4   6   3   2   8   1   10   15   11

1, 2, 3, 4, 6, 7, 8, 9, 10, 11, 15
Range = 15 − 1 = 14
Lower quartile = 3
Upper quartile = 10
Interquartile range = 10 − 3
= 7

Put the list in ascending order before trying to find the quartiles.

There are 11 values and $\frac{11}{4}$ = 2.75 so rounding up gives 3. Q_1 is therefore the third term when the numbers are put in ascending order.

1, 2, **3**, 4, 6, 7, 8, 9, 10, 11, 15

$\frac{3 \times 11}{4}$ = 8.25 so rounding up gives 9.

NB Although 8.25 is closer to 8 than 9 you always round up. Q_3 is therefore the ninth value when the numbers are put in ascending order.

This means 50% of the data lie within seven units.

Example 2

Rebecca records the number of CDs in the collections of students in her year.

The results are in the table opposite.

Find the interquartile range.

x	Number of students, f
35	3
36	17
37	29
38	34
39	12

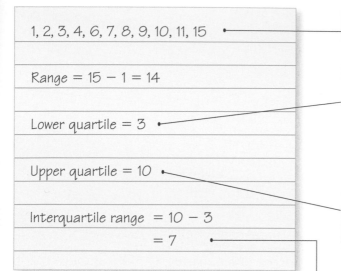

x	Number of students, f	Cumulative frequency
35	3	3
36	17	20
37	29	49
38	34	83
39	12	95

First calculate the cumulative frequency.

Lower quartile: $\frac{95}{4}$ = 23.75

which rounds up to 24.

Therefore Q_1 is the 24th value = 37

The 20th observation is 36 and the 49th term is 37 with every observation in between having a value of 37. Therefore the 24th value is 37.

Upper quartile: $\dfrac{3 \times 95}{4} = 71.25$

which rounds up to 72.

Therefore Q_3 is the 72nd value $= 38$

Interquartile range $= 38 - 37 = 1$ •————— 50% of the observations lie within one of each other.

Example 3

The length of time (to the nearest minute), spent on the internet each evening by a group of students is shown in the table below. Calculate the interquartile range.

Length of time spent on internet (minutes)	Number of students	Cumulative frequency
30–31	2	2
32–33	25	27
34–36	30	57
37–39	13	70

Lower quartile: $\dfrac{70}{4} = 17.5$ •————— Since this is grouped data we do not round up.

Using interpolation (see page 21)

This is the same method as for finding the median.

$\dfrac{Q_1 - 31.5}{33.5 - 31.5} = \dfrac{17.5 - 2}{27 - 2}$

$\dfrac{Q_1 - 31.5}{2} = \dfrac{15.5}{25}$

$Q_1 - 31.5 = \dfrac{15.5}{25} \times 2$

$Q_1 = 31.5 + 1.24$

$= 32.74$

Upper quartile: $\dfrac{3 \times 70}{4} = 52.5$

Using interpolation

33.5	Q_3	36.5
27	52.5	57

$$\frac{Q_3 - 33.5}{36.5 - 33.5} = \frac{52.5 - 27}{57 - 27}$$

$$\frac{Q_3 - 33.5}{3} = \frac{25.5}{30}$$

$$= 36.05$$

Interquartile range $= 36.05 - 32.74$

$$= 3.31$$

Exercise 3A

1 15 students do a mathematics test. Their marks are shown opposite.

7	4	9	7	6
10	12	11	3	8
5	9	8	7	3

  a Find the value of the median.

  b Find Q_1 and Q_3.

  c Work out the interquartile range.

2 A group of workers were asked to write down their weekly wage. The wages were:

£550   £400   £260   £320   £500   £450   £460   £480   £510   £490   £505

  a Work out the range for these wages.

  b Find Q_1 and Q_3.

  c Work out the interquartile range.

3 A superstore records the number of hours overtime worked by their employees in one particular week. The results are shown in the table.

  a Fill in the cumulative frequency column and work out how many employees the superstore had in that week.

  b Find Q_1 and Q_3.

  c Work out the interquartile range.

Number of hours	Frequency	Cumulative frequency
0	25	
1	10	
2	20	
3	10	
4	25	
5	10	

4 A moth trap was set every night for five weeks. The number of moths caught in the trap was recorded. The results are shown in the table.

Number of moths	Frequency
7	2
8	5
9	9
10	14
11	5

Find the interquartile range.

5 The weights of 31 Jersey cows were recorded to the nearest kilogram. The weights are shown in the table.

Weight of cattle (kg)	Frequency	Cumulative frequency
300–349	3	
350–399	6	
400–449	10	
450–499	7	
500–549	5	

a Complete the cumulative frequency column in the table.

b Find the lower quartile, Q_1.

c Find the upper quartile, Q_3.

d Find the interquartile range.

6 The number of visitors to a hospital in a week was recorded. The results are shown in the table.

Number of visitors	Frequency
500–1000	10
1000–1500	25
1500–2000	15
2000–2500	5
2500–3000	5

Giving your answers to the nearest whole number find:

a the lower quartile Q_1,　　**b** the upper quartile Q_3,　　**c** the interquartile range.

7 The lengths of a number of slow worms were measured, to the nearest mm. The results are shown in the table.

   a Work out how many slow worms were measured.

   b Find the interquartile range for the lengths of the slow worms.

Lengths of slow worms (mm)	Frequency
125–139	4
140–154	4
155–169	2
170–184	7
185–199	20
200–214	24
215–229	10

3.2 Percentiles split the data into 100 parts.

■ To calculate the xth **percentile**, P_x, you find the value of the $\frac{xn}{100}$th term.

■ The $n\%$ to $m\%$ **interpercentile range** $= P_m - P_n$

> The interquartile range is the 25% to 75% interpercentile range.

Example **4**

The height, in cm, of 70 seventeen year old boys were recorded as opposite:

Calculate

a the 90th percentile,

b the 10th percentile,

c the 10% to 90% interpercentile range.

Height	Number of students
150–160	4
160–170	21
170–180	32
180–190	9
190–200	4

   a P_{90} is the value of the $\frac{90}{100} \times 70 = 63$rd term

   Using interpolation

> This is the same method as for finding the median.

$$\frac{P_{90} - 180}{190 - 180} = \frac{63 - 57}{66 - 57}$$

$$P_{90} = 186.7 \text{ cm}$$

b P_{10} is the value of the $\dfrac{(10 \times 70)}{100} = $ 7th term

Using interpolation

$\dfrac{P_{10} - 160}{170 - 160} = \dfrac{(7 - 4)}{(25 - 4)}$

$P_{10} = 161.4$ cm

The 10th percentile is sometimes called the first decile (D_1).

c 10% to 90% interpercentile range

   = 186.7 cm − 161.4 cm

   = 25.3 cm

This means 90% − 10% = 80% of the data lie within 25.3 cm.

Exercise 3B

1 A gardener counted the peas in a number of pea pods. The results are shown in the table.

Number of peas	Frequency	Cumulative frequency
6	8	
7	12	
8	36	
9	18	
10	15	
11	10	

a Complete the cumulative frequency column.

b Calculate the 80th percentile.

c Calculate the 40th percentile

d Calculate the 65th percentile.

2 A shopkeeper goes to a clothes fair. He records the costs of jeans. The costs are shown in the table.

Cost of jeans (£'s)	Frequency	Cumulative frequency
10–15	11	
16–20	35	
21–25	34	
26–30	16	
31–35	10	
36–40	5	

  a Complete the cumulative frequency table.

  b Calculate P_{20}.

  c Calculate P_{80}.

  d Calculate the 20% to 80% interpercentile range.

3 The table shows the monthly income for a number of workers in a factory.
Calculate the 34% to 66% interpercentile range.

Monthly income (£'s)	Frequency
900–1000	3
1000–1100	24
1100–1200	28
1200–1300	15

4 A train travelled from Lancaster to Preston. The times, to the nearest minute, it took for the journey were recorded over a certain period. The times are shown in the table.

Time for journey (minutes)	15–16	17–18	19–20	21–22
Frequency	5	10	35	10

Calculate the 5% to 95% interpercentile range.

5 A roadside assistance firm kept a record over a week of the amount of time, in minutes, people were kept waiting for assistance in a particular part of the country. The time taken was from the time the phone call was received to the arrival of the breakdown mechanic. The times are shown below.

Time Waiting (minutes)	20–30	30–40	40–50	50–60	60–70
Frequency	6	10	18	13	2

  a Work out the number of people who called for assistance.

  b Calculate the 30th percentile.

  c Calculate the 65th percentile.

3.3 You can find the standard deviation and variance of discrete data.

The deviation of an observation x from the mean is given by $x - \bar{x}$, so one way of measuring the total dispersion (spread) of a set of data is to use the **variance.**

■ variance $= \dfrac{\sum(x - \bar{x})^2}{n}$

A more friendly version of this formula is

■ variance $= \dfrac{\sum x^2}{n} - \left(\dfrac{\sum x}{n}\right)^2$

> The following saying may help you remember this formula. Variance = 'mean of the squares minus the square of the mean'.

As variance is measured in units^2 you usually take the square root of the variance. The square root of the variance is called the **standard deviation**.

■ standard deviation $= \sqrt{\text{Variance}}$

The symbol σ is the symbol used for the standard deviation of the population and s is used when we estimate the standard deviation of the population (group being studied).

The notation used on most calculators where s and σ are not used is σ_n for σ and σ_{n-1} for s.

You should be aware that the formulas for s and σ differ slightly but in S1 you are not required to know the difference and will be using σ.

Example 5

The marks scored in a test by seven randomly selected students are

3   4   6   2   8   8   5

Find the variance of the marks of these seven students.

Mean $\bar{x} = \dfrac{36}{7}$

> Work in fractions to avoid rounding errors.

The deviation from the mean for each observation is

$-\dfrac{15}{7}$ $-\dfrac{8}{7}$ $\dfrac{6}{7}$ $-\dfrac{22}{7}$ $\dfrac{20}{7}$ $\dfrac{20}{7}$ $-\dfrac{1}{7}$

> Deviation $= x - \bar{x}$
> e.g. $3 - \dfrac{36}{7} = -\dfrac{15}{7}$

The deviations squared are

$\dfrac{225}{49}$ $\dfrac{64}{49}$ $\dfrac{36}{49}$ $\dfrac{484}{49}$ $\dfrac{400}{49}$ $\dfrac{400}{49}$ $\dfrac{1}{49}$

> (Deviation)$^2 = (x - \bar{x})^2$
> e.g. $\left(-\dfrac{15}{7}\right)^2 = \dfrac{225}{49}$

variance $= \dfrac{\frac{225 + 64 + 36 + 484 + 400 + 400 + 1}{49}}{7}$

$= 4.69$

> Variance $= \dfrac{\sum(x - \bar{x})^2}{n}$

Example 6

The marks gained in a test by seven randomly selected students are

3   4   6   2   8   8   5

Use the friendly formula to work out the variance and standard deviation of the marks of the seven students.

x	3	4	6	2	8	8	5
x^2	9	16	36	4	64	64	25

$\sum x = 3 + 4 + 6 + 2 + 8 + 8 + 5 = 36$

$\sum x^2 = 9 + 16 + 36 + 4 + 64 + 64 + 25 = 218$

variance, $\sigma^2 = \dfrac{218}{7} - \left(\dfrac{36}{7}\right)^2 = 4.69$

standard deviation, $\sigma = \sqrt{4.69} = 2.17$

> It is much quicker to use this friendly formula and less likely that rounding errors are made.

Exercise 3C

1 Given that for a variable x:

$\sum x = 24$ $\sum x^2 = 78$ $n = 8$

Find:

a The mean.   **b** The variance σ^2.   **c** The standard deviation σ.

2 Ten collie dogs are weighed (w kg). The following summary data for the weights are shown below:

$\sum w = 241$ $\sum w^2 = 5905$

Use this summary data to find the standard deviation of the collies' weights.

3 Eight students' heights (h cm) are measured. They are as follows:

165   170   190   180   175   185   176   184

a Work out the mean height of the students.

b Given $\sum h^2 = 254\,307$ work out the variance. Show all your working.

c Work out the standard deviation.

4 For a set of 10 numbers:

$\sum x = 50$ $\sum x^2 = 310$

For a set of 15 numbers :

$\sum x = 86$ $\sum x^2 = 568$

Find the mean and the standard deviation of the combined set of 25 numbers.

5 The number of members (m) in six scout groups was recorded. The summary statistics for these data are:

$$\sum m = 150 \qquad \sum m^2 = 3846$$

    a Work out the mean number of members in a scout group.

    b Work out the standard deviation of the number of members in the scout groups.

6 There are two routes for a worker to get to his office. Both the routes involve hold ups due to traffic lights. He records the time it takes over a series of six journeys for each route. The results are shown in the table.

Route 1	15	15	11	17	14	12
Route 2	11	14	17	15	16	11

    a Work out the mean time taken for each route.

    b Calculate the variance and the standard deviation of each of the two routes.

    c Using your answers to **a** and **b** suggest which route you would recommend. State your reason clearly.

3.4 **Calculation of the variance and standard deviation for a frequency table and a grouped frequency distribution where x is the mid-point of the class.**

If you let f stand for the frequency, then $n = \sum f$ and

$$\text{Variance} = \frac{\sum f(x - \bar{x})^2}{\sum f} \qquad \text{or} \qquad \frac{\sum fx^2}{\sum f} - \left(\frac{\sum fx}{\sum f}\right)^2$$

Example 7

Shamsa records the time spent out of school during the lunch hour to the nearest minute, x, of the female students in her year. The results are as follows.

x	Number of students, f
35	3
36	17
37	29
38	34
39	26

Calculate the standard deviation of the time spent out of school.

x	Number of students, f	fx	fx^2
35	3	$3 \times 35 = 105$	$3 \times 35^2 = 3675$
36	17	612	22 032
37	29	1073	39 701
38	34	1292	49 096
39	26	1014	39 546
total	109	4096	154 050

Add two columns to the table, fx and fx^2. Add the total at the bottom of the columns.

To calculate fx^2 calculate x^2 first then multiply by f.

$\sum fx^2 = 154\,050$

$\sum fx = 4096$

$\sum f = 109$

$\text{variance} = \dfrac{154\,050}{109} - \left(\dfrac{4096}{109}\right)^2$

$= 1.19805$

$\text{standard deviation} = \sqrt{1.19805}$

$= 1.09$

If you do not have a single value for each x in an interval you have to use the mid-point of the interval. This means that your answer will only be an estimate.

Example 8

Andy recorded the length, in minutes, of each telephone call he made for a month. The data is summarised in the table below.

Length of telephone call	Number of occasions
$0 < l \leqslant 5$	4
$5 < l \leqslant 10$	15
$10 < l \leqslant 15$	5
$15 < l \leqslant 20$	2
$20 < l \leqslant 60$	0
$60 < l \leqslant 70$	1

Calculate an estimate of the standard deviation of the length of telephone calls.

Length of telephone call	Number of occasions, f	Mid-point x	fx	fx^2
$0 < l \leqslant 5$	4	2.5	$4 \times 2.5 = 10$	$4 \times 6.25 = 25$
$5 < l \leqslant 10$	15	7.5	112.5	843.75
$10 < l \leqslant 15$	5	12.5	62.5	781.25
$15 < l \leqslant 20$	2	17.5	35	612.5
$20 < l \leqslant 60$	0	40	0	0
$60 < l \leqslant 70$	1	65	65	4225
total	27		285	6487.5

Use the mid-point of the interval for x.

$\sum fx^2 = 6487.5$

$\sum fx = 285$

$\sum f = 27$

$$\text{variance} = \frac{6487.5}{27} - \left(\frac{285}{27}\right)^2$$

$$= 128.85802$$

$$\text{standard deviation} = \sqrt{128.85802}$$

$$= 11.35$$

Exercise 3D

1 For a certain set of data:

$$\sum fx = 1975 \qquad \sum fx^2 = 52\,325 \qquad n = 100$$

Work out the variance for these data.

2 For a certain set of data

$$\sum fx = 264 \qquad \sum fx^2 = 6456 \qquad n = 12$$

Work out the standard deviation for these data.

3 Nahab asks the students in his year group how much pocket money they get per week. The results, rounded to the nearest pound, are shown in the table.

Number of £'s (x)	Number of students f	fx	fx^2
8	14		
9	8		
10	28		
11	15		
12	20		
Totals			

a Complete the table.

b Using the formula $\dfrac{\sum fx^2}{\sum f} - \left(\dfrac{\sum fx}{\sum f}\right)^2$ work out the variance for these data.

c Work out the standard deviation for these data.

4 In a student group, a record was kept of the number of days absence each student had over one particular term. The results are shown in the table.

Number days absent (x)	Number of students f	fx	fx^2
0	12		
1	20		
2	10		
3	7		
4	5		

a Complete the table.　　　　　　　　　b Calculate the variance for these data.

c Work out the standard deviation for these data.

5 A certain type of machine contained a part that tended to wear out after different amounts of time. The time it took for 50 of the parts to wear out was recorded. The results are shown in the table.

Lifetime in hours	Number of parts	Mid-point x	fx	fx^2
$5 < h \leqslant 10$	5			
$10 < h \leqslant 15$	14			
$15 < h \leqslant 20$	23			
$20 < h \leqslant 25$	6			
$25 < h \leqslant 30$	2			

a Complete the table.

b Calculate an estimate for the variance and the standard deviation for these data.

6 The heights (x cm) of a group of 100 women were recorded.

The summary data is as follows:

$$\sum fx = 17\,100 \qquad \sum fx^2 = 2\,926\,225$$

Work out the variance and the standard deviation.

3.5 **Use coding to make numbers easier to work with when the data values are large.**

- ■ Adding or subtracting numbers does not change the standard deviation of the data.

- ■ Multiplying or dividing the data by a number does affect the standard deviation.

- ■ To find the standard deviation of the original data, find the standard deviation of the coded data and either multiply this by what you divided the data by, or divide this by what you multiplied the data by.

Example 9

a Find the standard deviation of the following lengths.

x (mm)	110	120	130	140	150

b Use the given coding to find the standard deviation of the above data.

i $y = \dfrac{x}{10}$ **ii** $y = x - 100$ **iii** $y = \dfrac{x - 100}{10}$

---

a $\sum x = 650$

$\sum x^2 = 85\,500$

standard deviation $= \sqrt{\dfrac{85\,500}{5} - \left(\dfrac{650}{5}\right)^2}$

$= 14.1$

---

b i coded data 11, 12, 13, 14, 15

standard deviation of coded data

$= \sqrt{\dfrac{855}{5} - \left(\dfrac{65}{5}\right)^2}$

$= 1.41$

standard deviation of original data $= 14.1$

The standard deviation of the coded data is the standard deviation of the original data divided by 10.

To find the standard deviation of the original data multiply the standard deviation of the coded data by 10.

ii coded data   10, 20, 30, 40, 50

standard deviation of coded data

$$= \sqrt{\frac{5500}{5} - \left(\frac{150}{5}\right)^2}$$

$$= 14.1$$

standard deviation of original data $= 14.1$

> Adding or subtracting a number does not change the spread of the data. The range is still 40 so the standard deviation is the same for the coded data as for the original observations.

iii coded data   1, 2, 3, 4, 5

standard deviation of coded data

$$= \sqrt{\frac{55}{5} - \left(\frac{15}{5}\right)^2}$$

$$= 1.41$$

standard deviation of original data $= 14.1$

> Adding or subtracting a number does not change the spread of the data but dividing by 10 does make a difference. Multiply the standard deviation of the coded data by 10 to get the standard deviation of the original observations.

Example 10

Use the method of coding to estimate the standard deviation of the length of a telephone call to the nearest minute, l, given in the table below.

Length of telephone call	Number of occasions
$0 < l \leqslant 5$	4
$5 < l \leqslant 10$	15
$10 < l \leqslant 15$	5
$15 < l \leqslant 20$	2
$20 < l \leqslant 60$	0
$60 < l \leqslant 70$	1

Use the coding $y = \dfrac{l - 7.5}{5}$

The following table shows how to carry out the calculations.

Length of telephone call	Number of occasions, f	Mid-point l	$y = \dfrac{l - 7.5}{5}$	fy	fy^2
0–5	4	2.5	−1	−4	4
5–10	15	7.5	0	0	0
10–15	5	12.5	1	5	5
15–20	2	17.5	2	4	8
20–60	0	40	6.5	0	0
60–70	1	65	11.5	11.5	132.25
total	27			16.5	149.25

Use the frequency and the coded data.

Standard deviation of coded data $= \sqrt{\dfrac{149.25}{27} - \left(\dfrac{16.5}{27}\right)^2}$

$= 2.27$

Standard deviation of original data $= 2.27 \times 5$

$= 11.35$

Example 11

Data is coded using $y = \dfrac{x - 150}{50}$. The standard deviation of the coded data is 2.5.
Find the standard deviation of the original data.

Standard deviation $= 2.5 \times 50$

$= 125$

Exercise 3E

1 **a** Work out the standard deviation of the following data.

x 11   13   15   20   25

b Use the following coding to find the standard deviation of the data.
Show all the working.

i $x - 10$ **ii** $\dfrac{x}{10}$ **iii** $\dfrac{x - 3}{2}$

2 Use the following codings to find the standard deviation of the following weights, w.

| w | 210 | 260 | 310 | 360 | 410 |

i $\dfrac{w}{10}$ **ii** $w - 200$ **iii** $\dfrac{w - 10}{200}$

3 **a** The coding $y = \dfrac{x - 50}{28}$ gives a standard deviation of 0.01 for y.
Work out the standard deviation for x.

   b The coding $y = \dfrac{h}{15}$ produced a standard deviation for y of 0.045.
What is the standard deviation of h?

   c The coding $y = x - 14$ produced a standard deviation for y of 2.37.
What is the standard deviation of x?

   d The coding $y = \dfrac{s - 5}{10}$ produced a standard deviation for y of 0.65.
What is the standard deviation of s?

4 The coding $y = x - 40$ gives a standard deviation for y of 2.34.
Write down the standard deviation of x.

5 The lifetime, x, in hours, of 70 light bulbs is shown in the table.

Lifetime in hours	Number of light bulbs
20–22	3
22–24	12
24–26	40
26–28	10
28–30	5
total	70

Using the coding $y = \dfrac{x - 1}{20}$ estimate the standard deviation of the actual lifetime in hours of a light bulb.

6 The weekly income, i, of 100 women workers was recorded.
The data were coded using $y = \dfrac{i - 90}{100}$ and the following summations were obtained.

$$\sum y = 131, \quad \sum y^2 = 176.84$$

Work out an estimate for the standard deviation of the actual women workers' weekly income.

7 A meteorologist collected data on the annual rainfall, x mm, at six randomly selected places.
The data were coded using $s = 0.01x - 10$ and the following summations were obtained.

$$\sum s = 16.1, \quad \sum s^2 = 147.03$$

Work out an estimate for the standard deviation of the actual annual rainfall.

Mixed exercise 3F

1 For the set of numbers:

$$2 \quad 5 \quad 3 \quad 8 \quad 9 \quad 10 \quad 3 \quad 2 \quad 15 \quad 10 \quad 5 \quad 7 \quad 6 \quad 7 \quad 5$$

    a work out the median,

    b work out the lower quartile,

    c work out the upper quartile,

    d work out the interquartile range.

2 A frequency distribution is shown below

Class interval	1–20	21–40	41–60	61–80	81–100
Frequency (*f*)	5	10	15	12	8

Work out the interquartile range.

3 A frequency distribution is shown below.

Class interval	1–10	11–20	21–30	31–40	41–50
Frequency (*f*)	10	20	30	24	16

    a Work out the value of the 30th percentile.

    b Work out the value of the 70th percentile.

    c Calculate the 30% to 70% interpercentile range.

4 The heights(*h*) of 10 mountains were recorded to the nearest 10 m and are shown in the table below.

h	1010	1030	1050	1000	1020	1030	1030	1040	1020	1000

Use the coding $y = \dfrac{h - 10}{100}$ to find the standard deviation of the heights.

5 The times it took a random sample of runners to complete a race are summarised in the table.

    a Work out the lower quartile.

    b Work out the upper quartile.

    c Work out the interquartile range.

The mid-point of each class was represented by *x* and its corresponding frequency by *f* giving:

$$\sum fx = 3740 \qquad \sum fx^2 = 183\,040$$

    d Estimate the variance and standard deviation for these data.

Time taken (*t* minutes)	Frequency
20–29	5
30–39	10
40–49	36
50–59	20
60–69	9

6 20 birds were caught for ringing. Their wing spans (x) were measured to the nearest centimetre.

The following summary data was worked out:

$$\sum x = 316 \qquad \sum x^2 = 5078$$

a Work out the mean and the standard deviation of the wing spans of the 20 birds.

One more bird was caught. It had a wing span of 13 centimetres

b Without doing any further working say how you think this extra wing span will affect the mean wing span.

7 The heights of 50 clover flowers are summarised in the table.

Heights in mm (h)	Frequency
90–95	5
95–100	10
100–105	26
105–110	8
110–115	1

a Find Q_1.

b Find Q_2.

c Find the interquartile range.

d Use $\sum fx = 5075$ and $\sum fx^2 = 516\,112.5$ to find the standard deviation.

Summary of key points

1   The **range** of a set of data is the difference between the highest and lowest value in the set.

2   The **quartiles**, Q_1, Q_2, Q_3, Q_4, split the data into four parts. To calculate the **lower quartile, Q_1**, divide n by 4.

3   For discrete data for the lower quartile, Q_1, divide n by 4. To calculate the upper quartile, Q_3, divide n by 4 and multiply by 3. When the result is a whole number find the mid-point of the corresponding term and the term above. When the result is not a whole number round the number up and pick the corresponding term.

4   For continuous grouped data for Q_1, divide n by 4 and for Q_3 divide n by 4 and multiply by 3. Use interpolation to find the value of the corresponding term.

5   The **interquartile range** is $Q_3 - Q_1$.

6   **Percentiles** split the data into 100 parts.
   - **To calculate the xth percentile**, P_x, find the value of the $\frac{xn}{100}$th term.
   - The $n\%$ to $m\%$ interpercentile range $= P_m - P_n$.

7   The **deviation** of an observation x from the mean is given by $x - \bar{x}$.
   - **Variance** $= \dfrac{\sum(x - \bar{x})^2}{n} = \dfrac{\sum x^2}{n} - \left(\dfrac{\sum x}{n}\right)^2$
   - **Standard deviation** $= \sqrt{\textbf{Variance}}$

8   To work out the variance and standard deviation for a frequency table and a grouped frequency distribution where x is the mid-point of the class, let f stand for the frequency, then $n = \sum f$
   - Variance $= \dfrac{\sum f(x - \bar{x})^2}{\sum f}$ or $\dfrac{\sum fx^2}{\sum f} - \left(\dfrac{\sum fx}{\sum f}\right)^2$

9   When the data values are large you can use coding to make the numbers easier to work with. To find the standard deviation of the original data, find the standard deviation of the coded data and either multiply this by what you divided by, or divide this by what you multiplied by.

After completing this chapter you should be able to

- draw **stem and leaf** diagrams
- calculate **outliers**
- draw **box plots**
- draw **histograms**
- work out whether data are **skewed**
- compare sets of data

You will also know when to use the various measures of **location** and **dispersion**.

Representation of data

A traveller wishes to compare two different train companies which both run trains between town A and town B. What is the best way to represent and compare this data?

Which train is the best one to catch?

4.1 Drawing and interpreting stem and leaf diagrams.

- A **stem and leaf diagram** is used to order and present data given to two or three significant figures. Each number is first split into its **stem** and **leaf**. Take the number 42.

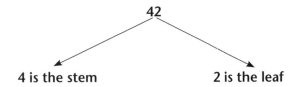

42

4 is the stem   2 is the leaf

- A stem and leaf keeps the detail of the data but can be time consuming to do.

- It enables the shape of the distribution of the data to be revealed, and quartiles can easily be found from the diagram.

- Two sets of data can be compared by using back-to-back stem and leaf diagrams.

Example 1

The blood glucose of 30 males is recorded. The results, in mmol/litre, are given below.

4.7	3.6	3.8	4.7	4.1	2.2	3.6	4.0	4.4	5.0	3.7	4.6	4.8	3.7	3.2
2.5	3.6	4.5	4.7	5.2	4.7	4.2	3.8	5.1	1.4	2.1	3.5	4.2	2.4	5.1

Draw a stem and leaf diagram to represent these data.

```
stem | leaf
  1  | 4
  2  | 2 5 1 4
  3  | 6 8 6 7 7 2 6 8 5
  4  | 7 7 1 0 4 6 8 5 7 7 2 2
  5  | 0 2 1 1
```

When drawing a stem and leaf diagram, first enter the data as it appears.

Remember to put in a key.

```
stem | leaf              Key: 2|1 means 2.1
  1  | 4
  2  | 1 2 4 5
  3  | 2 5 6 6 6 7 7 8 8
  4  | 0 1 2 2 4 5 6 7 7 7 7 8
  5  | 0 1 1 2
```

Once you have all the data you should put the leaves in ascending order. This will allow it to be used for calculations.

Try and keep the numbers in vertical columns.

This row contains all the numbers between 5.0 and 5.9.

Example 2

Find

a the mode,

b the lower quartile,

c the upper quartile,

d the median,

for the data in Example 1.

stem	leaf								Key: 2|1 means 2.1		
1	4										(1)
2	1	2	4	5							(4)
3	2	5	6	6	6	7	7	8	8		(9)
4	0	1	2	2	4	5	6	7	7	7 7 8	(12)
5	0	1	1	2							(4)

> This row contains all the numbers between 2.0 and 2.9

> This is the mode, there are 4 7s. The mode is 4.7.

> This is the number of pieces of data in the row.

a From the diagram you can see the mode is 4.7.

b Lower quartile: $\dfrac{30}{4} = 7.5$ so pick the 8th term = 3.6.

c Upper quartile: $\dfrac{3(30)}{4} = 22.5$ so pick the 23rd term = 4.7.

d Median: $\dfrac{(30)}{2} = 15$ so pick the 15.5th term = 4.05.

Example 3

John recorded the resting pulse rate for the 16 boys and 23 girls in his year at school.
The results were as follows

Girls

55	80	84	91	80
92	98	40	60	64
66	72	96	85	88
90	76	54	58	92
78	80	79		

Boys

80	60	91	65	67
59	75	46	72	71
74	57	64	60	50
68				

a Construct a back-to-back stem and leaf diagram to represent these data.

b Comment on your results.

a Unordered

		Girls			Boys	
		0	4	6		
	8 4 5	5	9 7 0			
	6 4 0	6	0 5 7 4 0 8			
	9 8 6 2	7	5 2 1 4			
0 8 5 0 4 0	8	0				
2 0 6 8 2 1	9	1				

Key: 0|4|6 means 40 for the girls and 46 for the boys

Ordered

	Girls			Boys	
	0	4	6		
	8 5 4	5	0 7 9		
	6 4 0	6	0 0 4 5 7 8		
	9 8 6 2	7	1 2 4 5		
8 5 4 0 0 0	8	0			
8 6 2 2 1 0	9	1			

Key: 0|4|6 means 40 for the girls and 46 for the boys

The lowest value always goes next to the stem when ordering.

b The back-to-back stem and leaf diagram shows that boys' pulse rate tends to be lower than the girls'.

Exercise 4A

1 A group of thirty college students was asked how many DVDs they had in their collection. The results are as follows.

12	25	34	17	12	18	29	34	45	6
15	9	25	23	29	22	20	32	15	15
19	12	26	27	27	32	35	42	26	25

Draw a stem and leaf diagram to represent these data.
  a Find the median.   **b** Find the lower quartile.   **c** Find the upper quartile.

2 The following stem and leaf diagram shows some information about the marks gained by a group of students in a statistics test.

stem	leaf								Key: 2|3 means 23 marks	
0	8	9								(2)
1	2	5	5	9						(4)
2	3	6	6	6	7					(5)
3	4	4	5	7	7	7	7	7	9	(9)
4	5	8	8	9						(4)

  a Work out how many students there were in the group.
  b Write down the highest mark.

c Write down the lowest mark.

d Write down how many students got 26 marks.

e Write down the modal mark.

f Find the median mark.

g Find the lower quartile.

h Find the upper quartile.

3 The number of laptops sold by a store was recorded each month for a period of 26 months. The results are shown in the stem and leaf diagram.

stem	leaf								Key: 1|8 means 18 laptops	
1	8									(1)
2	3	6	7	9	9					(5)
3	2	6	6	6	7	8	8			(7)
4	4	5	5	5	7	7	7	7	9	(9)
5	2	7	7	9						(4)

a Find the median.

b Find the lower quartile.

c Find the upper quartile.

d Work out the interquartile range.

e Write down the modal number of laptops sold.

4 A class of 16 boys and 13 girls did a Physics test. The test was marked out of 60. Their marks are shown below.

	Boys				Girls		
45	54	32	60	26	54	47	32
28	34	54	56	34	34	45	46
32	29	47	48	39	52	24	28
44	45	56	57	33			

a Draw a back-to-back stem and leaf diagram to represent these data.

b Comment on your results.

5 The following stem and leaf diagram shows the weekend earnings of a group of college students.

Males		Females	Key: 5|1|0 means £15 for males
8	0	6	and £10 for females
7 6 5	1	0 5 5 5 8 8	
9 9 9 8 6 6	2	5 5 8 8 9	
8 8 5 5 5	3	5 5	
8 5	4	0	

a Write down the number of male students and the number of female students.

b Write down the largest amount of money earned by the males.

c Comment on whether males or females earned the most in general.

4.2 An outlier is an extreme value.

An **outlier** is an extreme value that lies outside the overall pattern of the data.
The following rule is often used to identify outliers.

An outlier is any value, which is

■ **greater than the upper quartile + 1.5 × interquartile range**

or

■ **less than the lower quartile − 1.5 × interquartile range.**

Different rules are sometimes used and in the S1 examination you will be told what rule to apply.

Example 4

The blood glucose of 30 females is recorded. The results, in mmol/litre, are shown in the stem and leaf diagram below.

stem	leaf												Key: 2|1 means 2.1	
2	2	2	3	3	5	7								(6)
3	1	2	6	7	7	7	8	8	8	8	9	9	9	(13)
4	0	0	0	0	4	5	6	7	8					(9)
5	1	5												(2)

An outlier is an observation that falls
either 1.5 × interquartile range above the upper quartile
or 1.5 × interquartile range below the lower quartile.

a Find the quartiles.

b Find any outliers.

a Lower quartile: $\dfrac{30}{4} = 7.5$ so pick the eighth term $= 3.2$

Upper quartile: $\dfrac{3(30)}{4} = 22.5$ so pick the 23rd term $= 4.0$

Median: $\dfrac{(30)}{2} = 15$ so pick the 15.5th term $= 3.8$

b Interquartile range $= 4.0 - 3.2 = 0.8$

Outliers are values less than $3.2 - 1.5 \times 0.8 = 2$

and values greater than $4.0 + 1.5 \times 0.8 = 5.2$.

therefore 5.5 is an outlier.

When finding the quartiles from a stem and leaf diagram be careful to go through the data in ascending order.

Exercise 4B

1 Some data are collected. The lower quartile is 46 and the upper quartile is 68.

An outlier is an observation that falls either 1.5 × (interquartile range) above the upper quartile or 1.5 × (interquartile range) below the lower quartile.

Work out whether the following are outliers using this rule.

a 7

b 88

c 105

2 Male and female turtles were weighed in grams. For males, the lower quartile was 400 g and the upper quartile was 580 g. For females, the lower quartile was 260 g and the upper quartile was 340 g.

An outlier is an observation that falls either 1 × (interquartile range) above the upper quartile or 1 × (interquartile range) below the lower quartile.

a Which of these male turtle weights would be outliers?
   400 g   260 g   550 g   640 g

b Which of these female turtle weights would be outliers?
   170 g   300 g   340 g   440 g

c What is the largest size a male turtle can be without being an outlier?

4.3 A box plot can be drawn to represent important features of the data. It shows the quartiles maximum and minimum values and any outliers.

■ A **box plot** looks like this.

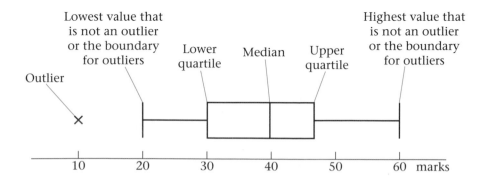

There is more than one method of labelling the whiskers when outliers are present. In this book we shall usually end the whisker at the next value in the data before the outlier. In some cases, where the full data are not available, you will have to use the boundary for outliers (see Example 6).

Example 5

The blood glucose of 30 females is recorded. The results, in mmol/litre, are shown in the stem and leaf diagram below.

stem	leaf													Key: 2|1 means 2.1	
2	2	2	3	3	5	7								(6)
3	1	2	6	7	7	7	8	8	8	8	9	9	9	(13)
4	0	0	0	0	4	5	6	7	8					(9)
5	1	5												(2)

Draw a box plot to represent the data.

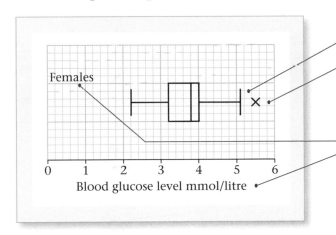

The outlier is marked with a cross. The highest value which is not an outlier is 5.1.

Always use a scale and label it. Remember to give your box plot a title.

Exercise 4C

1 A group of students did a test. The summary data is shown in the table below.

Lowest value	Lower quartile	Median	Upper quartile	Highest value
5	21	28	36	58

Given that there were no outliers draw a box plot to illustrate these data.

2 Here is a box plot of marks in an examination.

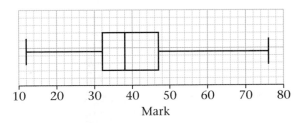

a Write down the upper and lower quartiles.

b Write down the median.

c Work out the interquartile range.

d Work out the range.

4.4 Using box plots to compare two sets of data.

Two sets of data can be **compared** using box plots.

Example 6

The blood glucose level of 30 males is recorded. The results, in mmol/litre, are summarised below.

Lower quartile   = 3.6
Upper quartile   = 4.7
Median   = 4.0
Lowest value   = 1.4
Highest value   = 5.2

An outlier is an observation that falls either $1.5 \times$ interquartile range above the upper quartile or $1.5 \times$ interquartile range below the lower quartile.

a Given that there is only one outlier, draw a box plot for these data on the same diagram as the box plot drawn in Example 5.

b Compare the blood glucose level for males and females.

a   Outliers are values less than
    $3.6 - 1.5 \times 1.1 = 1.95$
    and values greater than
    $4.7 + 1.5 \times 1.1 = 6.35$.
    There is 1 outlier which is 1.4

> The end of the whisker is plotted at the outlier boundary (in this case 1.95) as we do not know the actual figure.

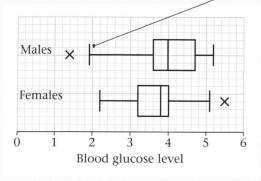

> When drawing two box plots use the same scale so they can be compared. Remember to give each a title and label the axis.

b   The median blood glucose for
    females is lower than the median
    blood glucose for males.
    The interquartile range (the width
    of the box) and range for blood
    glucose are smaller for the females.

> When comparing data you should compare a measure of location and a measure of spread. You should also write your interpretation in the context of the question.

Exercise 4D

1 A group of students took a statistics test. The summary data for the percentage mark gained by boys and by girls is shown in the box plots opposite.

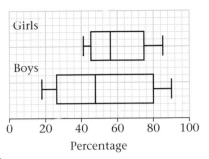

   a Write down the percentage mark which 75% of the girls scored more than.

   b State the name given to this value.

   c Compare and contrast the results of the boys and the girls.

2 Male and female turtles were weighed in grams. Their weights are summarised in the box plots opposite.

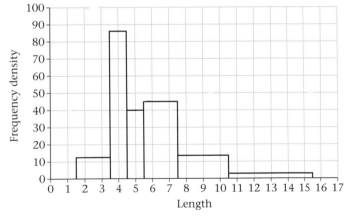

   a Compare and contrast the weights of the male and female turtles.

   b A turtle was found to be 330 grams in weight. State whether it is likely to be a male or a female. Give a reason for your answer.

   c Write down the size of the largest female turtle.

4.5 **Data can be represented by a histogram if they are continuous and are summarised in a group frequency distribution.**

■ A **histogram** gives a good picture of how data are distributed. It enables you to see a rough location, the general shape of the data and how spread out the data are.

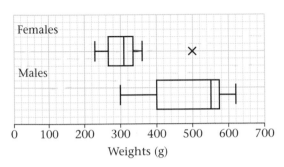

■ A histogram is similar to a **bar chart** but are two major differences

   • There are no gaps between the bars.

   • The **area** of the bar is proportional to the **frequency**.

■ To calculate the height of each bar (the **frequency density**) use the formula
Area of bar = $k \times$ frequency.

   $k = 1$ is the easiest value to use when drawing a histogram then

$$\text{frequency density} = \frac{\text{frequency}}{\text{class width}}$$

Example 7

A random sample of 200 students was asked how long it took them to complete their homework the previous night. The time was recorded and summarised in the table below.

Time (minutes)	25–30	30–35	35–40	40–50	50–80
Frequency	55	39	68	32	6

a Draw a histogram to represent the data.

b Estimate how many students took between 36 and 45 minutes to complete their homework.

a

Time (minutes)	Frequency	Class width	Frequency density
25–30	55	5	11
30–35	39	5	7.8
35–40	68	5	13.6
40–50	32	10	3.2
50–80	6	30	0.2

Frequency density $= \frac{55}{5}$
(height of bar) $= 11$

Class width $= 30 - 25$
$= 5$

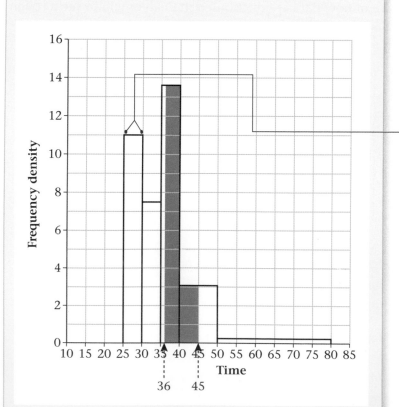

On the x-axis the bars use the class boundaries e.g. the first bar goes from 25 to 30.

Remember to label the axis. The vertical one is always frequency density.

b To estimate the number of students who spent between 36 and 45 minutes you need to find the area between 36 and 45

Shaded area

$= (40 - 36) \times 13.6 + (45 - 40) \times 3.2$

$= 70.4$ students

Example 8

The histogram shows the variable *t* which represents the time taken, in seconds, by a group of children to solve a puzzle. The shaded bar A represents a frequency of 78 children.

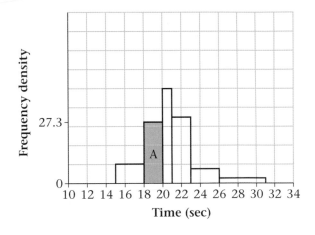

a Why should a histogram be used to represent these data?

b Write down the underlying feature associated with each of the bars in a histogram.

c What area on the histogram represents one child?

The total area under the histogram is 210 units².

d Find the total number of children in the group.

a Time is continuous.

b The area of the bar is proportional to the frequency.

c Area of bar A $= 2 \times 27.3$

$= 54.6$ units²

54.6 units² represents 78 children

one child is $\dfrac{54.6}{78} = 0.7$ units²

d $210 \div 0.7 = 300$ children

Exercise **4E**

1 The heights of a year group of children were measured. The data are summarised in the group frequency table.

Height (cm)	Frequency	Class width	Frequency density
135–144	40	10	
145–149	40	5	
150–154	75		
155–159	65		
160–174	60		

a Copy and complete the table.

b Draw a histogram for these data.

2 Some students take part in an obstacle race. The time it took each student to complete the race was noted. The results are shown in the histogram.

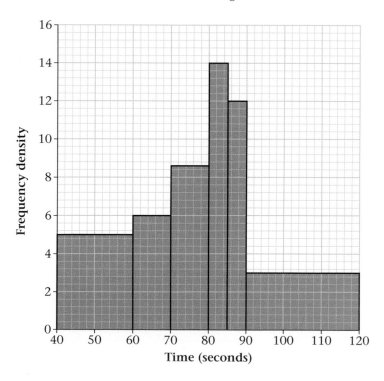

a Give a reason to justify the use of a histogram to represent these data.

The number of students who took between 60 and 70 seconds is 90.

b Find the number of students who took between 40 and 60 seconds.

c Find the number of students who took 80 seconds or less.

d Calculate the total number of students who took part in the race.

3 The time taken for each employee in a company to travel to work was recorded. The results are shown in the histogram.

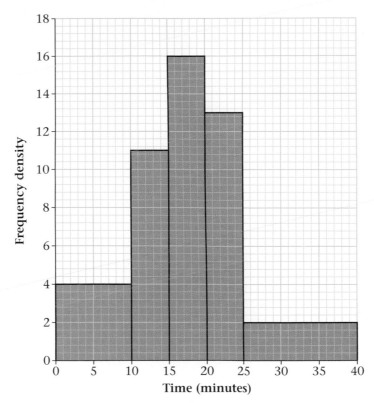

The number of employees who took less than 10 minutes to travel to work is 48.

a Find how many employees took less than 15 minutes to travel to work.

b Estimate how many employees took between 20 and 30 minutes to travel to work.

c Estimate how many employees took more than 30 minutes to travel to work.

4 A Fun Day committee at a local sports centre organised a throwing the cricket ball competition. The distance thrown by every competitor was recorded. The data were collected and are shown in the histogram. The number of competitors who threw less than 20 m was 40.

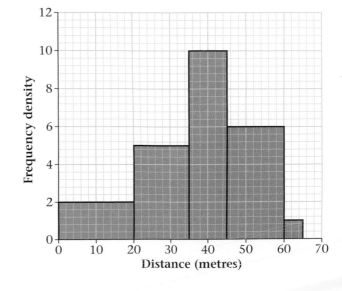

a Why is a histogram a suitable diagram to represent these data?

b How many people entered the competition?

c Estimate how many people threw between 30 and 40 metres.

d How many people threw between 45 and 65 metres?

e Estimate how many people threw less than 25 metres.

5 A farmer weighed a random sample of pigs. The weights were summarised in a grouped frequency table and represented by a histogram.

One of the classes in the grouped frequency distribution was 28–32 and its associated frequency was 32. On the histogram the height of the rectangle representing that class was 2 cm and the width was 2 cm.

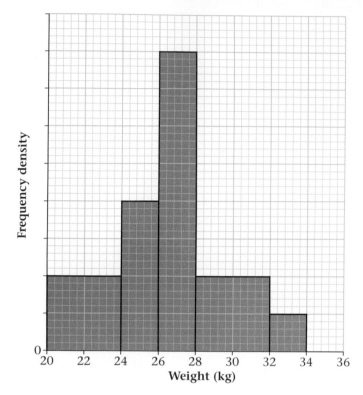

a Give a reason to justify the use of a histogram to represent these data.

b Write down the underlying feature associated with each of the bars in a histogram.

c Show that on this histogram each pig was represented by 0.125 cm².

d How many pigs did the farmer weigh altogether?

e Estimate the number of pigs that weighed between 25 and 29 kg.

4.6 **The shape (skewness) of a data set can be described using diagrams, measures of location and measures of spread.**

A distribution can be **symmetrical**, have **positive skew** or have **negative skew**.

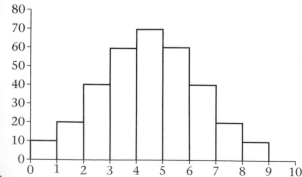

This distribution is said to be symmetric.

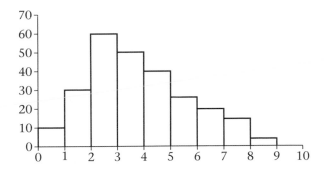

This distribution is said to have positive skew.

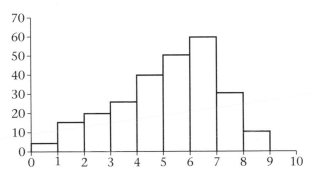

This distribution is said to have negative skew.

There are a number of ways of describing whether a distribution is skewed.

■ **You can use the quartiles.**

If $Q_2 - Q_1 = Q_3 - Q_2$ then the distribution is **symmetrical**.
If $Q_2 - Q_1 < Q_3 - Q_2$ then the distribution is **positively** skewed.
If $Q_2 - Q_1 > Q_3 - Q_2$ then the distribution is **negatively** skewed

■ **You can see the shape from box plots.**

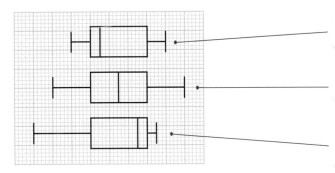

Positive skew – indicated by small distance between Q_1 and the median.

Symmetrical.

Negative skew – indicated by small distance between the median and Q_3.

■ **You can use the measures of location**

mode = median = mean   describes a distribution which is **symmetrical**.
mode < median < mean   describes a distribution with **positive skew**.
mode > median > mean   describes a distribution with **negative skew**.

■ **You can calculate** $\dfrac{3(\text{mean} - \text{median})}{\text{standard deviation}}$

This gives you a value and tells you how skewed the data are. The larger the number the greater the skew. The closer the number to zero the more symmetrical the data. A negative number means the data has negative skew. A positive number means the data has a positive skew.

Example 9

The following stem and leaf diagram shows the scores, x, obtained by a group of students in a test.

Score															Key: 6|1 means 61	
2	1	2	8													(3)
3	3	4	7	8	9											(5)
4	1	2	3	5	6	7	9									(7)
5	0	2	3	3	5	5	6	8	9	9						(10)
6	1	2	2	3	4	4	5	6	6	8	8	8	9	9		(14)
7	0	2	3	4	5	7	8	9								(8)
8	0	1	4													(3)

a Write down the modal score.

b Find the three quartiles for these data.

For these data, $\sum x = 2873$ and $\sum x^2 = 177\,353$.

c Calculate, to two decimal places, the mean and the standard deviation for these data.

d Calculate the value of $\dfrac{3(\text{mean} - \text{median})}{\text{standard deviation}}$ and comment on the skewness.

e Use two further methods to show that these data are negatively skewed.

---

a mode = 68

b $Q_1 \left(\dfrac{50}{4} = 12.5 \text{ therefore 13th term}\right) = 46$;

Using the method for discrete data.

$Q_2 \left(\dfrac{50}{2} = 25 \text{ therefore mean of the 25th and 26th term}\right) = 60$;

$Q_3 \left(3 \times \dfrac{50}{4} = 37.5 \text{ therefore 38th term}\right) = 69$

c $\mu = \dfrac{2873}{50} = 57.46$

$\sigma^2 = \dfrac{177\,353}{50} - (57.46)^2 = 245.4084$ ←————————— $\sigma^2 = \dfrac{\sum x^2}{n} - \left(\dfrac{\sum x}{n}\right)^2$

$\therefore \sigma = \sqrt{245.4084} = 15.7$

d $\dfrac{3(\text{mean} - \text{median})}{\text{standard deviation}} = \dfrac{3(57.46 - 60)}{15.7} = -0.486 \therefore$ negative skew

Remember to use accurate numbers in the calculation.

e $(Q_3 - Q_2) < (Q_2 - Q_1)$, i.e. $9 < 14 \Rightarrow$ negative skew

Mean < Median < Mode, i.e. $57.46 < 60 < 68 \Rightarrow$ negative skew

4.7 **Comparing the distributions of data sets and when to use each of the measures of spread and location to describe data sets. Interpreting these measures in the context of the question.**

A variety of measures is often used to compare data sets. When **comparing data sets** you should have the relevant information

1   a measure of location

2   a measure of spread

1 and 2 are the absolute minimum you should give.

3   skewness.

Bear in mind the following:

- The range gives a rough idea of the spread of the data but it is affected by extreme values. It is generally only used with small data groups together with either the median or mode

- The interquartile range (IQR) is not affected by extreme values and tells you how spread out the middle 50% of the observations are. The IQR is often used together with the median when the data are skewed.

- The mean and standard deviation are generally used when the data are fairly symmetrical, and the data size is not small.

Example 10

The values of daily sales, to the nearest £1, taken at a small shop last year are summarised in the table below.

Sales	Number of days
1–300	190
301–500	80
501–800	63
801–1100	27
1101–2000	5

a Draw a histogram to represent these data.

The shop owner wants to compare last year's sales with figures from another year.

b State whether the shop owner should use the median and the interquartile range or the mean and the standard deviation to compare daily sales. Give a reason for your answer.

a

Sales	Number of days	Class width	Frequency density
1–300	190	300	0.633
301–500	80	200	0.4
501–800	63	300	0.21
801–1100	27	300	0.09
1101–2000	5	900	0.006

> Use the class boundaries to find the class width, e.g. 300.5 − 0.5 = 300

> Frequency density = $\dfrac{\text{Frequency}}{\text{Class width}}$

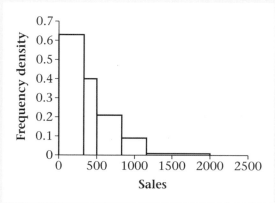

b Median and interquartile range because the data are skewed.

Example 11

A company runs two manufacturing lines A and B which make rods 2 cm in diameter. Random samples are taken from each of the lines A and B and the diameters measured. The results are summarised below.

	Mean diameter	Standard deviation of diameter
A	2	0.015
B	2	0.05

The company wishes to close one of the lines down. State which one you would recommend them to close. Given one reason for your answer.

> They should close down line B as the standard deviation is larger. In manufacturing you want the rods to all be the same so the smaller standard deviation is better.

> You must think about the context of the question. It is not always advantageous to have a small standard deviation as is seen in Example 12.

Example 12

The table summarises the marks, out of 75, for two different examinations.

	Mean mark	Standard deviation
Statistics	55	16
Mechanics	55	4

If you assume that the students are of mixed ability, state which of the papers is better for enabling you to set fair grade boundaries. Give a reason for your answer.

Statistics is the better paper as the standard deviation is bigger. This means the marks are more spread out so the grade boundaries are well spread out.

Exercise 4F

1 In a survey of the earnings of some sixth form students who did Saturday jobs the median wage was £36.50. The 75th percentile was £45.75 and the interquartile range was £30.50.

   Use the quartiles to describe the skewness of the distribution.

2 A group of estate agents recorded the time spent on the first meeting with a random sample of 120 of their clients. The times, to the nearest minute, are summarised in the table.

   a Calculate estimates of the mean and variance of the times.

   b By interpolation obtain estimates of the median and quartiles of the times spent with customers.

   One measure of skewness is found using $\dfrac{3(\text{mean} - \text{median})}{\text{standard deviation}}$.

   c Evaluate this measure and describe the skewness of these data.

   The estate agents are undecided whether to use the median and quartiles, or the mean and standard deviation to summarise these data.

   d State, giving a reason, which you would recommend them to use.

Time	Number of clients
10–15	2
15–20	5
20–25	17
25–30	38
30–35	29
35–45	25
45–80	4
Total	120

3 The following stem and leaf diagram summarises the wing length, to the nearest mm, of a random sample of 67 owl moths.

	Wing length												Key: 5|0 means 50
5	0 0 0 1 1 2 2 3 3 3 4 4												(12)
5	5 5 6 6 6 7 8 8 9 9												(10)
6	0 1 1 1 3 3 4 4 4 4												(10)
6	5 5 6 7 8 9 9												(7)
7	1 1 2 2 3 3												(6)
7	5 7 9 9												(4)
8	1 1 1 2 2 3 3 4												(8)
8	7 8 9												(3)
9	0 1 1 2												(4)
9	5 7 9												(3)

a Write down the mode of these data.

b Find the median and quartiles of these data.

c On graph paper, construct a box plot to represent these data.

d Comment on the skewness of the distribution.

e Calculate the mean and standard deviation of these data.

f Use a further method to show that these data are skewed.

g State, giving a reason, which of **b** or **e** you would recommend using to summarise the data in the table.

4 A TV company wishes to appeal to a wider range of viewers. They decide to purchase a programme from another channel. They have the option of buying one of two programmes. The company collects information from a sample of viewers for each programme. The results are summarised in the table.

	Mean age	Standard deviation of age
Programme 1	50	5
Programme 2	50	10

State which programme the company should buy to increase the range of their viewers. Give a reason for your answer.

Mixed exercise 4G

1 Jason and Perdita decided to go for a touring holiday on the continent for the whole of July. They recorded the number of kilometres they travelled each day. The data are summarised in the stem and leaf diagram below.

stem	leaf	Key: 15|5 means 155 kilometres
15	5	
16	4 8 9	
17	3 5 7 8 8 8 9 9 9	
18	4 4 5 5 8	
19	2 3 4 5 5 6	
20	4 7 8 9	
21	1 2	
22	6	

a Find Q_1, Q_2, and Q_3

Outliers are values that lie outside $Q_1 - 1.5(Q_3 - Q_1)$ and $Q_3 + 1.5(Q_3 - Q_1)$.

b Find any outliers.

c Draw a box plot of these data.

d Comment on the skewness of the distribution.

2 Sophie and Jack do a survey every day for three weeks. Sophie counts the number of pedal cycles using Market Street. Jack counts the number of pedal cycles using Strand Road. The data they collected are summarised in the back-to-back stem and leaf diagram.

Sophie	Stem	Jack	
9 9 7 5	0	6 6	Key: 5|0|6 means Sophie counts
7 6 5 3 3 2 2 2 1 1	1	1 1 5	5 cycles and Jack counts
5 3 3 2 2	2	1 2 2 2 3 7 7 8 9	6 cycles
2 1	3	2 3 4 7 7 8	
	4	2	

a Write down the modal number of pedal cycles using Strand Road.

The quartiles for these data are summarised in the table below.

	Sophie	Jack
Lower quartile	X	21
Median	13	Y
Upper quartile	Z	33

b Find the values for X, Y and Z.

c Write down the road you think has the most pedal cycles travelling along it overall. Give a reason for your answer.

3 Shop A and Shop B both sell mobile phones. They recorded how many they sold each day over a long period of time. The data they collected are represented in the box plots.

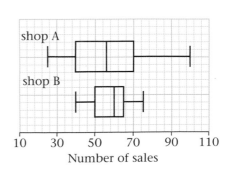

a Shop B says that for 50% of the days they sold 60 or more phones a day. State whether or not this is a true statement. Give a reason for your answer.

b Shop A says that for 75% of the days they sold 40 or more phones a day. State whether or not this is a true statement. Give a reason for your answer.

c Compare and contrast the two box plots.

d Write down the shop you think had the most consistent sales per day. Explain the reason for your choice.

4 Fell runners from the Esk Club and the Irt Club were keen to see which club had the fastest runners overall. They decided that all the members from both clubs would take part in a fell run. The time each runner took to complete the run was recorded.

The results are summarised in the box plot.

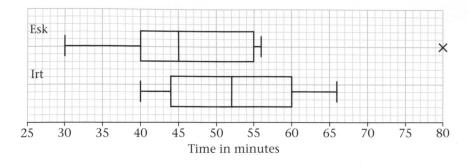

Time in minutes

a Write down the time by which 50% of the Esk Club runners had completed the run.

b Write down the time by which 75% of the Irt Club runners had completed the run.

c Explain what is meant by the cross (✗) on the Esk Club box plot.

d Compare and contrast these two box plots.

e Comment on the skewness of the two box plots.

f What conclusions can you draw from this information about which club has the fastest runners?

5 The histogram shows the time taken by a group of 58 girls to run a measured distance.

a Work out the number of girls who took longer than 56 seconds.

b Estimate the number of girls who took between 52 and 55 seconds.

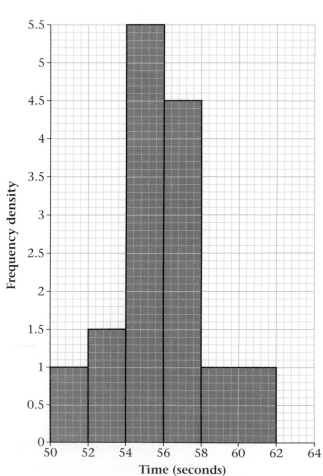

6 The table gives the distances travelled to school, in km, of the population of children in a particular region of the United Kingdom.

Distance, km	0–1	1–2	2–3	3–5	5–10	10 and over
Number	2565	1784	1170	756	630	135

A histogram of this data was drawn with distance along the horizontal axis. A bar of horizontal width 1.5 cm and height 5.7 cm represented the 0–1 km group.
Find the widths and heights, in cm to one decimal place, of the bars representing the following groups:

a 2–3, **b** 5–10.

7 The labelling on bags of garden compost indicates that the bags weigh 20 kg.
The weights of a random sample of 50 bags are summarised in the table opposite.

Weight in kg	Frequency
14.6–14.8	1
14.8–18.0	0
18.0–18.5	5
18.5–20.0	6
20.0–20.2	22
20.2–20.4	15
20.4–21.0	1

a On graph paper, draw a histogram of these data.

b Estimate the mean and standard deviation of the weight of a bag of compost.

[You may use $\sum fy = 988.85$, $\sum fy^2 = 19\,602.84$]

c Using linear interpolation, estimate the median.

One coefficient of skewness is given by

$$\frac{3(\text{mean} - \text{median})}{\text{standard deviation}}.$$

d Evaluate this coefficient for the above data.

e Comment on the skewness of the distribution of the weights of bags of compost.

8 The number of bags of potato crisps sold per day in a bar was recorded over a two-week period. The results are shown below.

20 15 10 30 33 40 5 11 13 20 25 42 31 17

a Calculate the mean of these data.

b Draw a stem and leaf diagram to represent these data.

c Find the median and the quartiles of these data.

An outlier is an observation that falls either 1.5 × (interquartile range) above the upper quartile or 1.5 × (interquartile range) below the lower quartile.

d Determine whether or not any items of data are outliers.

e On graph paper draw a box plot to represent these data. Show your scale clearly.

f Comment on the skewness of the distribution of bags of crisps sold per day. Justify your answer.

Summary of key points

1   A stem and leaf diagram is used to order and present data.

2   A stem and leaf diagram reveals the shape of the data and enables quartiles to be found.

3   Two sets of data can be compared using back-to-back stem and leaf diagrams.

4   An outlier is an extreme value. You will be told what rule to apply to identify outliers.

5   A box plot represents important features of the data. It shows quartiles, maximum and minimum values and any outliers.

6   Box plots can be used to compare two sets of data.

7   If data are continuous and summarised in a group frequency distribution, the data can be represented by a histogram.

8   Diagrams, measures of location and measures of spread can be used to describe the shape (skewness) of a data set.

9   You can describe whether a distribution is skewed using
   • quartiles
   • shape from box plots
   • measures of location
   • formula $\dfrac{3(\text{mean} - \text{median})}{\text{standard deviation}}$ (larger value means greater skew).

After completing this chapter you should be able to

- understand common terms used in probability and solve simple probability questions
- use set notation and Venn diagrams to solve probability problems with two or three events
- use given formulae to find probabilities
- understand conditional probability and how it relates to tree diagrams
- understand mutually exclusive and independent events including sampling with and without replacement
- find probabilities using arrangements.

Probability

Probability is a key part of mathematics. It plays an important role in the money markets, in weather reporting, in quantum theory and many other areas where statistics are applied to the real world.

5.1 Understanding the vocabulary used in probability.

- If you want to predict the chance of something happening in the future, you use probability.
  - An **experiment** is a repeatable process that gives rise to a number of **outcomes**.
  - An **event** is a collection (or set) of *one or more* outcomes.
  - A **sample space** is the set of all possible outcomes of an experiment.

- Events and sample spaces are sets and set notation is used to describe them. We use upper case letters to denote events. The probability of an event is the chance that the event will occur as a result of an experiment.
  - Where outcomes are equally likely the **probability of an event** is the number of outcomes in the event divided by the total number of possible outcomes in the sample space.
  - An impossible event has probability 0 and an event that is certain has probability 1.
  - As all events have probabilities between impossible (0) and certain (1), then probabilities are usually written as a fraction, a decimal or sometimes as a percentage. In S1 we will write probabilities as fractions or decimals.

Example 1

A fair die has each face numbered 1 to 6. The die is thrown once and the number landing face up is recorded.

a Find the probability of the die landing with the number 5 face up.

b Find the probability of throwing an odd number.

a There are six faces and one has the number 5 on it, so the probability of landing with the number 5 face up is $\frac{1}{6}$ or $0.1\dot{6}$.	The **sample space** has six outcomes since any one of the numbers 1, 2, 3, 4, 5 or 6 could land face up. The die is fair, so all outcomes are equally likely.
	There is one outcome in the event '5 is face up' divided by the total of six possible outcomes.
b The odd numbered faces are 1, 3 and 5, so the probability of throwing an odd number is $\frac{3}{6} = \frac{1}{2}$ or 0.5.	The **event** 'throwing an odd number' has three outcomes i.e. 1, 3 or 5 face up.
	There are three outcomes divided by the total of six possible outcomes.

Example 2

Two fair spinners each have four faces numbered 1 to 4. The two spinners are thrown together and the sum of the numbers indicated on each spinner is recorded.

Find the probability of the spinners indicating a sum of

a exactly 5,

b more than 5.

Draw two axes to indicate the values on each spinner.

The sample space is represented by the 16 points on the diagram.

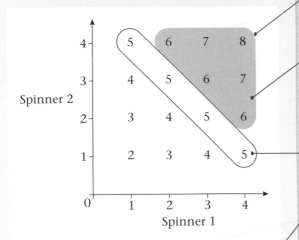

There are 4 × 4 = 16 points. Each of these points is equally likely as the spinners are fair.

There are 4 outcomes for this part. They are 1 + 4, 2 + 3, 3 + 2 or 4 + 1.

You can use *F* to represent the event 'the sum of the scores is exactly 5'.

P( ) is short for 'probability of'. The answer P(*F*) can also be written as 0.25.

a   There are four 5s and 16 outcomes altogether, so

$$P(F) = \frac{4}{16} = \frac{1}{4}$$

b   P(More than 5) $= \frac{6}{16} = \frac{3}{8}$

There are six sums more than 5 for this part (shaded blue). They form the top right hand corner of the diagram: 2 + 4, 3 + 4, 4 + 4, 3 + 3, 4 + 3, 4 + 2. The answer can also be written as 0.375.

Exercise 5A

For each of the following experiments, identify the sample space and find the probability of the event specified.

1   Throwing a six sided die once and recording if the number face up is odd or even. Find the probability of an even number landing face up.

2   Tossing two coins. Find the probability of the same outcome on each coin.

3   A card is drawn from a pack of 52 playing cards. Find the probability that the card is a heart.

4   A card is drawn from a pack of 52 playing cards, its suit is recorded then it is replaced and another card is drawn. Find the probability of drawing a spade and a club in any order.

5   Throwing a die and tossing a coin. Find the probability of a head and a 6.

6   Throwing two six sided dice and recording the product of the values on the two sides that are uppermost. Find the probability of the answer being greater than or equal to 24.

5.2 You can solve probability problems by drawing a Venn diagram.

- Events can be represented graphically by a Venn diagram. Venn diagrams are named after Hull born mathematician John Venn (1834–1923).

- We use set notation to identify different areas on a Venn diagram.

- You can write numbers of outcomes in the Venn diagram or the probabilities of the events to help to solve problems.

- You can use Venn diagrams to solve probability problems for three events (see Example 5).

Venn diagrams

A rectangle represents the sample space and it contains closed curves that represent events.

For events A and B in a sample space S:

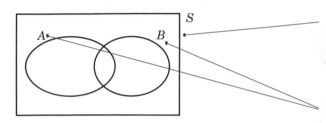

A rectangle labelled S represents the sample space. It includes all the possible outcomes of an experiment so the probability of the sample space is 1 or P(S) = 1.

A closed curve labelled A represents all the outcomes of event A and similarly for B.

1. The event $A \cap B$

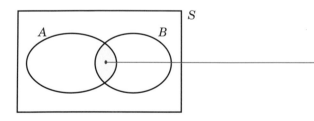

The overlap between event A and event B is written $A \cap B$.

This is A intersection B and represents the event that **both** A **and** B occur.

2. The event $A \cup B$

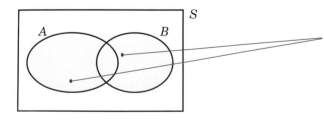

The union of event A and event B is written $A \cup B$.

This is A union B and represents the event that **either** A **or** B occurs or **both** occur.

3. The event A'

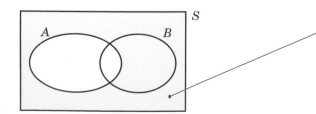

We are also interested in events **not** occurring. For example, A' is the complement of A and is the event 'A does not occur'.

The probability of the complement of an event is easier to work out if you remember that the probability of the sample space is 1.

So P(A') = 1 − P(A).

Example 3

A card is selected at random from a pack of 52 playing cards. Let A be the event that the card is an ace and D the event that the card is a diamond. Find these.

a $P(A \cap D)$

b $P(A \cup D)$

c $P(A')$

d $P(A' \cap D)$

Cards are selected at random so there are 52 equally likely outcomes in total, one for each card.

You start at the intersection of the sets and work outwards by subtracting.

Draw a Venn Diagram:

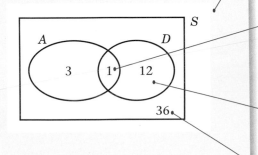

There are 13 diamonds and 4 aces but there is only one card that is in both A and D. So there is one outcome in the intersection namely the ace of diamonds.

There are 13 diamonds, 4 aces and 1 ace of diamonds. This leaves 12 (= 13 − 1) cards that are diamonds but not aces and 3 (= 4 − 1) cards that are aces but not diamonds.

There are 52 cards and (3 + 12 + 1) = 16 counted in A or D, so that leaves 52 − 16 = 36 cards to make up the rest of S.

a $A \cap D$ is the event 'the card chosen is the ace of diamonds'.

$$P(A \cap D) = \frac{1}{52}$$

There is one outcome in $A \cap D$ and 52 outcomes in S so probability is $\frac{1}{52}$.

b $A \cup D$ is the event 'the card chosen is an ace or a diamond or the ace of diamonds'

$$P(A \cup D) = \frac{16}{52} = \frac{4}{13}$$

There are (3 + 1 + 12) = 16 outcomes in $A \cup D$ and 52 outcomes in S.

c A' is the event 'the card chosen is not an ace'

$$P(A') = \frac{48}{52} = \frac{12}{13}$$

From the Venn diagram you can add up all values **outside** A: 12 + 36 = 48.

N.B. It is best to give final answers as fractions in their lowest terms.

d $A' \cap D$ is the event 'the card chosen is not an ace and is a diamond'

$$P(A' \cap D) = \frac{12}{52} = \frac{3}{13}$$

From the Venn diagram, this is where everything outside A overlaps with D i.e.12.

Example 4

In a class of 30 students, 7 are in the choir, 5 are in the school band and 2 students are in the choir and the band. A student is chosen at random from the class.

a Draw a Venn diagram to represent this information.

Find the probability that

b the student is not in the band,

c the student is not in the choir nor in the band.

a

> There are 30 students in the sample space. A student is chosen at random so all outcomes are equally likely.

> Put the number in both the choir and the band in $B \cap C$.

> Then the 2 can be subtracted from each of the numbers of students in the Band, $5 - 2 = 3$ and then the Choir, $7 - 2 = 5$.

> The final region represents the events in the sample space that are not in C or B: $30 - (3 + 2 + 5) = 20$.

b A student not in the band is B'

$P(B') = 1 - P(B)$

$\qquad = 1 - \dfrac{5}{30}$

$\qquad = \dfrac{25}{30} = \dfrac{5}{6}$

> Probability of B is 5 outcomes of the 30 outcomes in the sample space.

c P(the student is not in the choir or the band)

$= \dfrac{20}{30} = \dfrac{2}{3}$

> The Venn diagram shows that there are $3 + 2 + 5 = 10$ students in total that are in the choir or the band, so that leaves 20 that are not in the choir or the band.

Example 5

A vet surveys 100 of her clients. She finds that

25 own dogs	15 own dogs and cats	11 own dogs and tropical fish
53 own cats	10 own cats and tropical fish	7 own dogs, cats and tropical fish
40 own tropical fish		

A client is chosen at random.

a Draw a Venn diagram to represent this information.

Find the probability that the client

b owns dogs only,

c does not own tropical fish

d does not own dogs, cats or tropical fish.

In this example probabilities are used in the Venn diagram but it could be solved by using the number of outcomes as in the previous two examples.

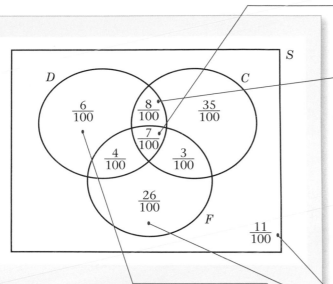

Start with $\frac{7}{100}$ in the intersection of all three events.

Work outwards to the intersections.

$\frac{15}{100} - \frac{7}{100} = \frac{8}{100}$

$\frac{10}{100} - \frac{7}{100} = \frac{3}{100}$

$\frac{11}{100} - \frac{7}{100} = \frac{4}{100}$

Each of 'dogs only', 'cats only' and 'tropical fish only' can be worked out by further subtractions:

$\frac{25}{100} - \frac{(8 + 7 + 4)}{100} = \frac{6}{100}$ for 'dogs only'

$\frac{40}{100} - \frac{(4 + 7 + 3)}{100} = \frac{26}{100}$ for 'fish only'

$\frac{53}{100} - \frac{(8 + 7 + 3)}{100} = \frac{35}{100}$ for 'cats only'

a P(client owns dogs only)

$= \frac{25 - (8 + 7 + 4)}{100} = \frac{6}{100} = \frac{3}{50}$

As the probability of the sample space is 1, the final area is

$1 - \frac{(26 + 4 + 7 + 3 + 6 + 8 + 35)}{100} = \frac{11}{100}$

b P(client does not own tropical fish)

$= \frac{100 - 40}{100} = \frac{60}{100} = \frac{3}{5}$

This is the value on the Venn diagram.

c P(client does not own any of these three types of pet) $= \frac{11}{100}$

This is the value on the Venn diagram outside *D*, *C* or *F*.

Exercise 5B

1 A card is chosen at random from a pack of 52 playing cards. *C* is the event 'the card chosen is a club' and *K* is the event 'the card chosen is a King'. Find these.

a P(*K*) **b** P(*C*) **c** P(*C* ∩ *K*)

d P(*C* ∪ *K*) **e** P(*C*′) **f** (*K*′ ∩ *C*)

2 There are 25 students in a certain tutor group at Philips College. There are 16 students in the tutor group studying German, 14 studying French and six students studying both French and German.

Find the probability that a randomly chosen student in the tutor group

a studies French,

b studies French and German,

c studies French but not German,

d does not study French or German.

3 On a firing range, a rifleman has two attempts to hit a target. The probability of hitting the target with the first shot is 0.2 and the probability of hitting with the second shot is 0.3. The probability of hitting the target with both shots is 0.1.

Find the probability of

a missing the target with both shots,

b hitting with the first shot and missing with the second.

4 Of all the households in the UK, 40% have a plasma TV and 50% have a laptop computer. There are 25% of households that have both a plasma TV and a laptop. Find the probability that a household chosen at random has either a plasma TV or a laptop computer but not both.

5 There are 125 diners in a restaurant who were surveyed to find out if they had ordered garlic bread, beer or cheesecake.

> 15 diners had ordered all three items
> 43 diners had ordered garlic bread
> 40 diners had ordered beer
> 44 diners had ordered cheesecake
> 20 had ordered beer and cheesecake
> 26 had ordered garlic bread and cheesecake
> 25 had ordered garlic bread and beer.

A diner is chosen at random. Find the probability that the diner ordered

a all three items,

b beer but not cheesecake and not garlic bread,

c garlic bread and beer but not cheesecake,

d none of these items.

6 A group of 275 people at a music festival were asked if they play guitar, piano or drums.

> one person plays all three instruments
> 65 people play guitar and piano
> 10 people play piano and drums
> 30 people play guitar and drums
> 15 people play piano only
> 20 people play guitar only
> 35 people play drums only

a Draw a Venn diagram to represent this information.

A festival goer is chosen at random from the group.
Find the probability that the person chosen

b plays piano

c plays at least two of guitar, piano or drums

d plays exactly one of the instruments

e plays none of the instruments.

5.3 Using formulae to solve problems.

There is a formula we can use for two events that links the probability of the union and the probability of the intersection.

If $P(A) = a$ and $P(B) = b$

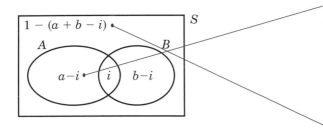

The probability of the intersection, $P(A \cap B)$, is i.

Subtract this probability from a and b and write the probabilities on the Venn diagram as shown.

The probability of $A \cup B$ is
$$P(A \cup B) = (a - i) + (b - i) + i$$
$$= a + b - i$$

If $P(S) = 1$ the area outside the closed curves has probability $1 - (a + b - i)$.

This gives us the formula

$$P(A \cup B) = (a - i) + (b - i) + i$$
$$= a + b - i$$

■ **Addition Rule** $P(A \cup B) = P(A) + P(B) - P(A \cap B)$

It is useful to remember that we can rearrange this if we want to find the intersection

$$P(A \cap B) = P(A) + P(B) - P(A \cup B)$$

We also know from Section 5.2 that $P(A') = 1 - P(A)$
These are given in the formula booklet.

Many probability problems are best solved by using the addition rule formula to find $P(A \cap B)$ and then drawing a Venn diagram to help find any other probabilities.

Example 6

A and B are two events and $P(A) = 0.6$, $P(B) = 0.7$ and $P(A \cup B) = 0.9$.
Find

a $P(A \cap B)$, **b** $P(A')$,

c $P(A' \cup B)$, **d** $P(A' \cap B)$.

a $P(A \cap B) = P(A) + P(B) - P(A \cup B)$
$P(A \cap B) = 0.6 + 0.7 - 0.9$
$P(A \cap B) = 0.4$

Use the formula and substitute in the probabilities.

b $P(A') = 1 - P(A)$
$P(A') = 1 - 0.6 = 0.4$

Now we know the probability of the intersection, we can draw a Venn diagram to help us.

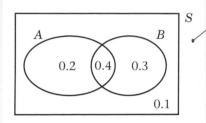

Write 0.4 in the intersection found in **a** then subtract to find the other values.

Use the Venn diagram to find the probability.

$A' \cup B$ is 'not A' together with B.

So we add up both the regions outside A (0.3 + 0.1) and the remaining region in B (0.4).

c $P(A' \cup B) = (0.3 + 0.1) + 0.4 = 0.8$

Now we know the union from **c**, we can use the addition formula and replace A with A'.

d $P(A' \cap B) = P(A') + P(B) - P(A' \cup B)$
$P(A' \cap B) = 0.4 + 0.7 - 0.8 = 0.3$

Of course you can simply identify the region $A' \cap B$ on your diagram as both outside A and inside B.

You can then simply write down the answer $P(A' \cap B) = 0.3$.

Exercise **5C**

1 A and B are two events and $P(A) = 0.5$, $P(B) = 0.2$ and $P(A \cap B) = 0.1$.
Find

a $P(A \cup B)$,
b $P(B')$,
c $P(A \cap B')$,
d $P(A \cup B')$.

2 A and C are two events and $P(A) = 0.4$, $P(B) = 0.5$ and $P(A \cup B) = 0.6$.
Find

a $P(A \cap B)$,
b $P(A')$,
c $P(A \cup B')$,
d $P(A' \cup B)$.

3 C and D are two events and $P(D) = 0.4$, $P(C \cap D) = 0.15$ and $P(C' \cap D') = 0.1$.
Find

a $P(C' \cap D)$,
b $P(C \cap D')$,
c $P(C)$,
d $P(C' \cap D')$.

4 There are two events T and Q where $P(T) = P(Q) = 3P(T \cap Q)$ and $P(T \cup Q) = 0.75$.
Find

a $P(T \cap Q)$,
b $P(T)$,
c $P(Q')$,
d $P(T' \cap Q')$,
e $P(T \cap Q')$.

5 A survey of all the households in the town of Bury was carried out. The survey showed that 70% have a freezer and 20% have a dishwasher and 80% have either a dishwasher or a freezer or both appliances. Find the probability that a randomly chosen household in Bury has both appliances.

6 The probability that a child in a school has blue eyes is 0.27 and the probability they have blonde hair is 0.35. The probability that the child will have blonde hair or blue eyes or both is 0.45. A child is chosen at random from the school. Find the probability that the child has

a blonde hair and blue eyes,

b blonde hair but not blue eyes,

c neither feature.

7 A patient going in to a doctor's waiting room reads *Hiya* Magazine with probability 0.6 and *Dakor* Magazine with probability 0.4. The probability that the patient reads either one or both of the magazines is 0.7. Find the probability that the patient reads
a both magazines, **b** *Hiya* Magazine only.

5.4 Solving problems using conditional probability.

The probability of an event B may be different if you know that a dependent event A has already occurred. Consider the Venn diagram below taken from Section 5.3.

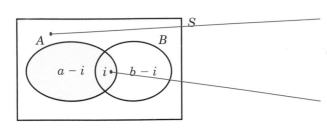

> The event A has happened, so we are considering probabilities inside this closed curve. You can think of it as the underlying sample space S 'shrunk' to the set A.

> We are looking for the probability of B given A has happened so we divide i by P(A) to rescale the probability of event A to 1. So
> P(B given A) = $\frac{i}{a}$.

So the probability of 'B given A' $= \dfrac{P(B \cap A)}{P(A)}$

The probability of B given A, written $P(B|A)$, is called the conditional probability of B given A

and so: $P(B|A) = \dfrac{P(B \cap A)}{P(A)}$

This also gives us a useful formula for the probability of the intersection.

■ **Multiplication Rule $P(B \cap A) = P(B|A) \times P(A)$**

We can also use $P(A \cap B) = P(A|B) \times P(B)$ when A and B are swapped.

Using the multiplication rule to find the probability of the intersection of two events and then drawing a Venn diagram is often the most efficient way of solving some problems. (See Example 8.)

Example 7

Two fair spinners each have four faces numbered 1 to 4. The two spinners are thrown together and the sum of the numbers indicated on each spinner is recorded.

Given that at least one spinner lands on a 3, find the probability of the spinners indicating a sum of exactly 5.

There is a clue in the question that we are looking for a conditional probability as the question uses the word 'given'.

Let A stand for the event 'at least one spinner lands on a 3' and B the event 'the spinners indicate a sum of exactly 5'.

Draw two axes to indicate the values on each spinner.

The outcomes in event A are highlighted in yellow in the diagram below.

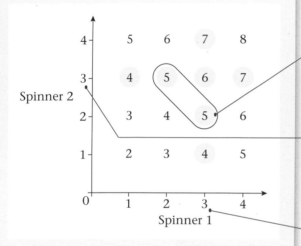

We know that one of the spinners lands on a 3. This reduces the number of outcomes where the sum is 5 to the two 5s shaded yellow.

There are seven outcomes altogether where one of the spinners lands on a 3 (shown circled), so the probability of at least one 3 is $\frac{7}{16}$, i.e $P(A) = \frac{7}{16}$.

METHOD 1: Using the formula.

$$P(B|A) = \frac{P(B \cap A)}{P(A)}$$

P(sum of 5 | at least one 3)

$$= \frac{P(\text{sum of 5 and at least one 3})}{P(\text{at least one 3})}$$

$$= \frac{\frac{2}{16}}{\frac{7}{16}}$$

$$= \frac{2}{7}$$

There are two 5s shaded in the diagram and 16 outcomes in the sample space.

METHOD 2: Using the 'shrunken' sample space.

As at least one spinner lands on a 3, this 'shrinks' the size of the sample space:

P(sum of 5 | at least one 3) $= \frac{2}{7}$

There are seven equally likely circled outcomes. Only two of these have a total of 5.

Example 8

C and D are two events such that $P(C) = 0.2$, $P(D) = 0.6$ and $P(C|D) = 0.3$.
Find

a $P(D|C)$,   **b** $P(C' \cap D')$,   **c** $P(C' \cap D)$.

Multiplication Rule •————————————

$P(C \cap D) = P(C|D) \times P(D)$

$\qquad = 0.3 \times 0.6 = 0.18$

> This is $P(A \cap B) = P(A|B) \times P(B)$ but with C and D instead.

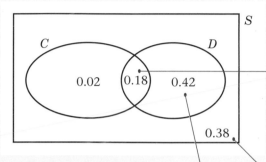

> Start with the intersection and work outwards by subtracting. This will help to work out any other probabilities in the question.

a $P(D|C) = \dfrac{P(D \cap C)}{P(C)}$

$\qquad = \dfrac{0.18}{0.2} = 0.9$

> This is the intersection of 'not C' and 'not D'.

b $P(C' \cap D') = 0.38$ •

c $P(C' \cap D) = 0.42$ •————————————

> This is the intersection of 'not C' with D.

Example 9

Let A and B be events such that $P(A) = \frac{3}{10}$, $P(B) = \frac{2}{5}$ and $P(B \cup A) = \frac{1}{2}$.
Find

a $P(B|A)$,   **b** $P(B|A')$,   **c** $P(A)P(B|A) + P(A')P(B|A')$.

Comment on your answer to part **c**.

a $P(B \cap A) = P(B) + P(A) - P(B \cup A)$ •

$\qquad = \dfrac{2}{5} + \dfrac{3}{10} - \dfrac{1}{2} = \dfrac{1}{5}$

> We use the addition rule to work out the intersection first.

$P(B|A) \quad = \dfrac{P(B \cap A)}{P(A)}$

> Then we can use the formula for conditional probability.

$\qquad = \dfrac{\frac{1}{5}}{\frac{3}{10}} = \dfrac{2}{3}$

b

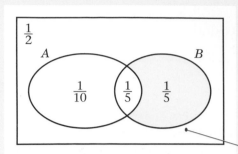

$P(B \cap A') = P(B) - P(B \cap A)$

$= \dfrac{2}{5} - \dfrac{1}{5}$

$= \dfrac{1}{5}$

> *B* overlapping with 'not *A*' is *B* without the intersection.

$P(B|A') = \dfrac{P(B \cap A')}{P(A')}$

> Then we can use the formula for conditional probability.

$= \dfrac{\dfrac{1}{5}}{1 - \dfrac{3}{10}} = \dfrac{2}{7}$

c $P(A)P(B|A) + P(A')P(B|A')$

$= \dfrac{3}{10} \times \dfrac{2}{3} + \dfrac{7}{10} \times \dfrac{2}{7} = \dfrac{2}{5}$

This is the same as the probability of event *B* illustrated in the Venn Diagram below.

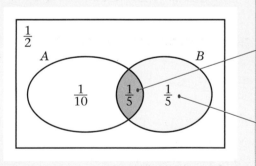

> $P(A)P(B|A) = P(B \cap A)$ by the Multiplication Rule.

> $P(A')P(B|A') = P(B \cap A')$ by the Multiplication Rule.

Exercise **5D**

1 A card is drawn at random from a pack of 52 playing cards. Given that the card is a diamond, find the probability that the card is an ace.

2 Two coins are flipped and the results are recorded. Given that one coin lands on a head, find the probability of

a two heads, **b** a head and a tail.

3 Two fair dice are thrown and the product of the numbers on the dice is recorded. Given that one die lands on 2, find the probability that the product on the dice is

    a exactly 6,   **b** more than 5.

4 A and B are two events such that $P(A) = 0.6$, $P(B) = 0.5$ and $P(A \cap B) = 0.4$, find

    a $P(A \cup B)$,   **b** $P(B|A)$,   **c** $P(A|B)$,   **d** $P(A|B')$.

5 A and B are two events such that $P(A) = 0.4$, $P(B) = 0.5$ and $P(A|B) = 0.4$, find

    a $P(B|A)$,   **b** $P(A' \cap B')$,   **c** $P(A' \cap B)$.

6 Let A and B be events such that $P(A) = \frac{1}{4}$, $P(B) = \frac{1}{2}$ and $P(A \cup B) = \frac{3}{5}$. Find

    a $P(A|B)$,   **b** $P(A' \cap B)$,   **c** $P(A' \cap B')$.

7 C and D are two events and $P(C|D) = \frac{1}{3}$, $P(C|D') = \frac{1}{5}$ and $P(D) = \frac{1}{4}$, find

    a $P(C \cap D)$,   **b** $P(C \cap D')$,   **c** $P(C)$,

    d $P(D|C)$,   **e** $P(D'|C)$,   **f** $P(D'|C')$.

5.5 Conditional probabilities can be represented on a tree diagram.

Some questions can be answered using either a Venn diagram or a **tree diagram** (see Example 11).

Example 10

The turnout of spectators at a motor rally is dependent upon the weather. On a rainy day the probability of a big turnout is 0.4, but if it does not rain, the probability of a big turnout increases to 0.9. The weather forecast gives a probability of 0.75 that it will rain on the day of the race.

a Draw a tree diagram to represent this information.

Find the probability that

b there is a big turnout and it rains,   **c** there is a big turnout.

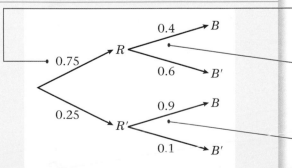

a   There are two events: R is the event 'it rains', B is the event 'there is a big turnout'.

First branches deal with 'rain' and the second branches deal with 'turnout'.

$P(R) = 0.75$ and $P(R') = 0.25$ are written on the first pair of branches.

Given it does rain, we are told the probability of a big turnout i.e. $P(B|R) = 0.4$, so we put 0.4 on the second tier after R. We put $1 - 0.4 = 0.6$ on the branch below.

Similarly $P(B|R') = 0.9$, so we put 0.9 on the second tier after R'. The other branch is $1 - 0.9 = 0.1$

CHAPTER 6

The multiplication rule is used to work out the probabilities.

b $P(B \cap R) = P(B|R)P(R)$

$= 0.4 \times 0.75$

$= 0.3$

This is the same as 0.75×0.4 from the top branches on the diagram. We multiply along the branches to event R then event B.

c It can either rain and there is a big turnout or it does not rain and there is a big turnout.

$P(B) = 0.3 + P(B|R')P(R')$

$= 0.3 + 0.25 \times 0.9$

$= 0.525$

This is the answer to **a** plus the probability found by multiplying along the branches from R' to B.

Remember, moving **along** the branches of a tree diagram you **multiply** probabilities, and moving **between** branches you **add** the probabilities. The probabilities on the second tier of branches are conditional upon the first tier.

Example 11

A and B are two events such that $P(A|B) = 0.1$, $P(A|B') = 0.6$ and $P(B) = 0.3$.
Find

a $P(A \cap B)$,   **b** $P(A \cap B')$,   **c** $P(A)$,   **d** $P(B|A)$,   **e** $P(B|A')$.

a by Multiplication Rule

$P(A \cap B) = P(A|B) \times P(B)$

$= 0.1 \times 0.3$

$= 0.03$

b by Multiplication Rule

$P(A \cap B') = P(A|B') \times P(B')$

$= 0.6 \times 0.7$

$= 0.42$

We use the Multiplication Rule first to find the intersection, then draw a Venn diagram to help with the rest of the question.

c

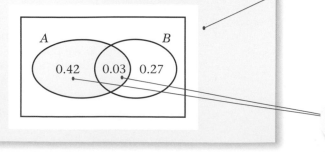

We can see from the Venn diagram that $P(A) = P(A \cap B') + P(A \cap B)$.

From Venn diagram

$$P(A) = 0.42 + 0.03$$
$$= 0.45$$

d $P(B|A) = \dfrac{P(B \cap A)}{P(A)}$

$$= \dfrac{0.03}{0.45} = 0.0\dot{6}$$

e $P(B|A') = \dfrac{P(B \cap A')}{P(A')}$

> We can use $P(A') = 1 - P(A)$ to find the denominator
> i.e. $P(A') = 1 - 0.45 = 0.55$.

$$= \dfrac{0.27}{0.55} = 0.4\dot{9}\dot{0}$$

Alternatively using a tree diagram:

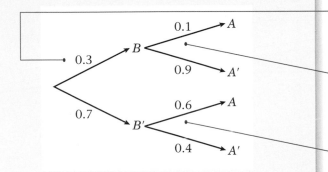

> The first branches give $P(B) = 0.3$ and therefore the $P(B') = 0.7$.

> The conditional probability $P(A|B) = 0.1$ is on the second branch and the other branch giving $P(A'|B)$ is simply $1 - P(A|B) = 1 - 0.1 = 0.9$.

> Similarly $P(A|B') = 0.6$, so we put 0.6 on the second tier after B'. The other branch is $1 - 0.6 = 0.4$.

a $P(A \cap B) = 0.3 \times 0.1 = 0.03$

> Multiply along the top branches.

b $P(A \cap B') = P(B' \cap A) = 0.7 \times 0.6$
$$= 0.42$$

> Simply multiply along the appropriate branches.

c $P(A) = P(A \cap B) + P(A \cap B') = 0.45$

> Simply add the previous two answers. Remember multiply along branches and add between them.

The solutions to d and e are the same. The formulae for conditional probability are used and then previous answers substituted.

Example 12

A bag contains seven green beads and five blue beads. A bead is taken from the bag at random, the colour is recorded and it is not replaced. A second bead is then taken from the bag and its colour recorded. Find the probability that one bead is green and another bead is blue.

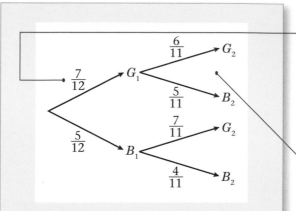

Initially there are 12 beads in the sample space, seven green and five blue. The first branch represents the event 'first bead chosen is green', labelled G_1, and the branch below is 'first bead chosen is blue' labelled B_1.

There are only 11 beads left in the bag after the first bead is chosen.

For the second tier of branches, the probabilities are conditional upon the first choice of bead. On this branch, the first bead chosen is green, so there are only six green beads left in the bag.

P(one bead is green and another is blue)

= P(first is green then second blue)

+ P(first is blue then second is green)

Multiply along the branches

$$= \frac{7}{12} \times \frac{5}{11} + \frac{5}{12} \times \frac{7}{11}$$

Add between branches.

$$= \frac{35}{66}$$

Notice the effect of not replacing the beads in Example 12. The different effects of replacement and no replacement will be discussed in Section 5.6.

Exercise 5E

1 A bag contains five red and four blue tokens. A token is chosen at random, the colour recorded and the token is not replaced. A second token is chosen and the colour recorded. Find the probability that

a the second token is red given the first token is blue,

b the second token is blue given the first token is red,

c both tokens chosen are blue,

d one red token and one blue token are chosen.

2 A box of 24 chocolates contains 10 dark and 14 milk chocolates. Linda chooses a chocolate at random and eats it, followed by another one. Fine the probability that Linda eats

   a two dark chocolates,

   b one dark and one milk chocolate.

3 Jean always goes to work by bus or takes a taxi. If one day she goes to work by bus, the probability she goes to work by taxi the next day is 0.4. If one day she goes to work by taxi, the probability she goes to work by bus the next day is 0.7.

   Given that Jean takes the bus to work on Monday, find the probability that she takes a taxi to work on Wednesday.

4 Sue has two coins. One is fair, with a head on one side and a tail on the other. The second is a trick coin and has a tail on both sides. Sue picks up one of the coins at random and flips it.

   a Find the probability that it lands heads up.

   b Given that it lands tails up, find the probability that she picked up the fair coin.

5 A contestant on a quiz show is asked to choose one of three doors. Behind one of the doors is the star prize of a sports car, but behind each of the other two doors there is a toy car.

   The contestant chooses one of the three doors.

   The host then opens one of the remaining two doors and reveals a toy car. The host then asks the contestant if they want to stick with their first choice or switch to the other unopened door.

   State what you would recommend the contestant to do in order to have the greatest probability of winning the sports car. Show your working clearly.

5.6 Mutually exclusive and independent events.

When events have no outcomes in common, they are **mutually exclusive**.

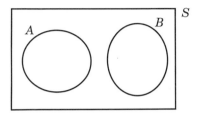

When A and B are mutually exclusive, the intersection of A and B is empty, so $P(A \cap B) = 0$.

We can use $P(A \cup B) = P(A) + P(B) - P(A \cap B)$ with $P(A \cap B) = 0$ to give:

■ The **Addition Rule** applied to **mutually exclusive** events:

   $P(A \cup B) = P(A) + P(B)$

When one event has no effect on another, they are **independent**. Therefore if A and B are independent, the probability of A happening is the same whether or not B has happened $P(A|B) = P(A)$.

We can use the formula for conditional probability to give $\dfrac{P(A \cap B)}{P(B)} = P(A)$

Rearranging gives:

■ The **Multiplication Rule** applied to **independent events**:

$$P(A \cap B) = P(A) \times P(B)$$

Example 13

Events A and B are mutually exclusive and $P(A) = 0.2$ and $P(B) = 0.4$.

Find

a $P(A \cup B)$,

b $P(A \cap B')$,

c $P(A' \cap B')$.

a

A and B are mutually exclusive so closed curves do not intersect and $P(A \cap B) = 0$.

$$P(A \cup B) = P(A) + P(B)$$

A and B are mutually exclusive so $P(A \cap B) = 0$.

$$= 0.2 + 0.4 = 0.6$$

b $P(A \cap B') = P(A)$

As there is no intersection between A and B, 'not B' overlaps with the whole of A.

$$= 0.2$$

c $P(A' \cap B') = 1 - P(A \cup B)$

The intersection of 'not A' and 'not B' is S without the union of A and B. Remember $P(S) = 1$.

$$= 1 - 0.6 = 0.4$$

Example 14

Events A and B are independent and $P(A) = \frac{1}{3}$ and $P(B) = \frac{1}{5}$.

Find

a $P(A \cap B)$,   **b** $P(A \cap B')$,   **c** $P(A' \cap B')$.

a $P(A \cap B) = P(A) \times P(B)$

$\qquad = \frac{1}{3} \times \frac{1}{5} = \frac{1}{15}$

A and B are independent, so we can use the simplified multiplication rule.

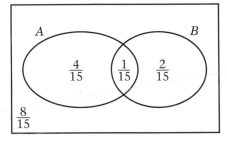

We can now draw a Venn diagram to help us with the rest of the question.

b $P(A \cap B') = P(A) - P(A \cap B)$

$\qquad = \frac{1}{3} - \frac{1}{15} = \frac{4}{15}$

*The intersection of A and 'not B' is A without the intersection of A with B. We use the answer to **a** or simply write down the answer from the diagram.*

Alternatively if A and B are independent, then so are A and B'

$P(A \cap B') = P(A) \times P(B')$

$\qquad = \frac{1}{3} \times \frac{4}{5} = \frac{4}{15}$ as before.

c $P(A \cup B) = P(A) + P(B) - P(A \cap B)$

$\qquad = \frac{1}{3} + \frac{1}{5} - \frac{1}{15} = \frac{7}{15}$

$P(A' \cap B') = 1 - P(A \cup B)$

$\qquad = 1 - \frac{7}{15} = \frac{8}{15}$

*We work out the union first from the Addition Rule. We use the answer to **a** again or simply add the probabilities from the diagram.*

The final answer of course can simply be written down from the diagram.

Alternatively if A and B are independent, then so are A' and B':

$P(A' \cap B') = P(A') \times P(B')$

$\qquad = \frac{2}{3} \times \frac{4}{5} = \frac{8}{15}$ as before.

Use $P(A') = 1 - \frac{1}{3} = \frac{2}{3}$ and $P(B') = 1 - \frac{1}{5} = \frac{4}{5}$.

Example 15

A red die and a blue die are rolled and the outcome on each die is recorded.

A is the event 'outcome on the red die is 3', B is the event 'outcome on the blue die is 3', C is the event ' the sum of the outcomes on each die is 5' and D is the event 'the outcome on each die is the same'.

Show that

a A and B are independent.

b C and D are mutually exclusive.

a We draw a diagram to represent the sample space with 1 to 6 for the red die along the bottom and 1 to 6 for the blue die on the left hand side.

Each loop contains 6 outcomes and they overlap at 'double 3'.

$P(A) = \dfrac{6}{36} = \dfrac{1}{6}, P(B) = \dfrac{6}{36} = \dfrac{1}{6}$

$P(A \cap B) = \dfrac{1}{36}$

There are $6 \times 6 = 36$ equally likely outcomes.

$P(A) \times P(B) = \dfrac{1}{6} \times \dfrac{1}{6}$

$= \dfrac{1}{36}$

Start by multiplying the two probabilities and show that they give the same answer as the probability of the intersection.

$= P(A \cap B)$

Therefore A and B are independent.

b

D has 6 outcomes:
(1, 1)
(2, 2)
(3, 3)
(4, 4)
(5, 5)
(6, 6)

C has 4 outcomes:
(4 + 1)
(3 + 2)
(2 + 3)
(1 + 4)

From the table, D has 6 outcomes and C has 4 outcomes.

Events C and D do not have outcomes in common, as the same value on each die cannot add up to 5 as it is an odd number.

Therefore C and D are mutually exclusive.

Example 16

In Section 5.5, Example 12 a tree diagram was used where the beads were taken 'without replacement'. In this example the beads are taken 'with replacement'.

A bag contains seven green beads and five blue beads. A bead is taken from the bag at random, the colour is recorded and the bead is replaced. A second bead is then taken from the bag and its colour recorded.

a Find the probability that one bead is green and another bead is blue.

b Show that the event 'the first bead is green' and the event 'the second bead is green' are independent.

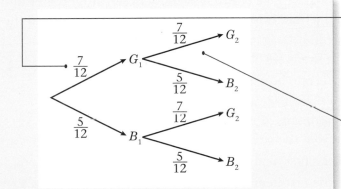

Initially there are 12 beads in the sample space, seven green and five blue. The first branch represents the event 'first bead chosen is green', labelled G_1, and the branch below is 'first bead chosen is blue' labelled B_1.

There are still 12 beads in the bag as the bead is replaced.

a P(one bead is green and the other bead is blue)

$= P(G_1 \cap B_2) + P(B_1 \cap G_2)$

$= \dfrac{7}{12} \times \dfrac{5}{12} + \dfrac{5}{12} \times \dfrac{7}{12}$

$= \dfrac{35}{72}$

On the second tier of branches the probabilities remain the same as the bead is **replaced**.

b $P(G_1) = \dfrac{7}{12}$

$P(G_2) = \dfrac{7}{12} \times \dfrac{7}{12} + \dfrac{5}{12} \times \dfrac{7}{12} = \dfrac{7}{12}$

$P(G_1 \cap G_2) = \dfrac{7}{12} \times \dfrac{7}{12}$

$= P(G_1) \times P(G_2)$ from above.

The event 'second bead is green' has two routes on the tree diagram; green first, green second or blue first, green second.

$G_1 \cap G_2$ is the event green first, green second from the tree diagram.

Therefore the event 'the first bead is green' and the event 'the second bead is green' are independent events. This is a feature of experiments 'with replacement'.

Exercise 5F

1 Event A and event B are mutually exclusive and $P(A) = 0.2$, $P(B) = 0.5$.
  a Draw a Venn diagram to represent these two events.
  b Find $P(A \cup B)$.
  c Find $P(A' \cap B')$.

2 Two events A and B are independent and $P(A) = \frac{1}{4}$ and $P(B) = \frac{1}{5}$.
  Find
  a $P(A \cap B)$,   **b** $P(A \cap B')$,   **c** $P(A' \cap B')$.

3 Q and R are two events such that $P(Q) = 0.2$, $P(R) = 0.4$ and $P(Q' \cap R) = 0.4$. Find
  a the relationship between Q and R,   **b** $P(Q \cup R)$,   **c** $P(Q' \cap R')$.

4 Two fair dice are rolled and the result on each die is recorded. Show that the event 'the sum of the scores on the dice is 4' and ' both dice land on the same number' are *not* mutually exclusive.

5 A bag contains three red beads and five blue beads. A bead is chosen at random from the bag, the colour is recorded and the bead is replaced. A second bead is chosen and the colour recorded.
  a Find the probability that both beads are blue.
  b Find the probability that the second bead is blue.

6 A box contains 24 electrical components of which four are known to be defective.
  Two items are taken at random from the box. Find the probability of selecting
  a two defective components if the first item is replaced before choosing the second item,
  b two defective components if the first item is not replaced,
  c one defective component and one fully functioning component if the first item is not replaced.

7 A bag contains one red, two blue and three green tokens. One token is chosen at random, the colour is recorded and the token replaced. A second token is then chosen and the colour recorded.
  a Draw a tree diagram showing the possible outcomes.
  Find the probability of choosing
  b two tokens of the same colour,
  c two tokens that are different colours.

8 Paul and Gill decide to play a board game. The probability that Paul wins the game is 0.25 and the probability that Gill wins is 0.3. They decide to play three games. Given that the results of successive games are independent, find the probability that
  a Paul wins three games in a row,   **b** all games are drawn,
  c Gill wins two games and Paul wins one game,   **d** each player wins just one game each.

5.7 Solving probability problems.

Drawing a diagram or other type of visual representation is a good problem-solving technique. It is often a sensible starting point for solving all kinds of probability problems.

Example 17

A local council consists of 100 elected members. There are 45 Labour, 25 Democrat and 30 Conservative councillors. A council delegation of three members is to be chosen at random from the council. Find the probability that

a the delegation has a representative from each political party,

b two members of the delegation are Labour and one is Conservative.

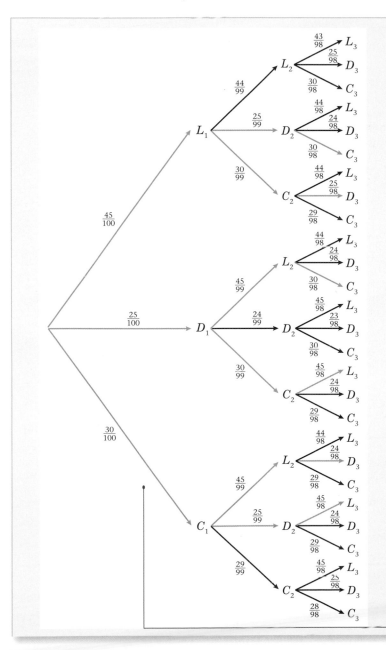

For the delegation to have a representative from each party, we need to find all the routes through the tree diagram that have one *L*, one *D* and one *C*.

Any of the three branches can be chosen at the start, this leaves two after the first choice then there is only one left at the end.

Therefore there are $3 \times 2 \times 1 = 6$ routes altogether. These are orange on the tree diagram.

Every route has 100, 99 and 98 on the bottom of each fraction. Every route has 45, 25 and 30 on the top of each fraction.

a P(the delegation has a representative from each political party)

$$= 6 \times \frac{45}{100} \times \frac{30}{99} \times \frac{25}{98}$$

$$= \frac{225}{1078}$$

> Multiply the fractions for each route and then add together all six routes.

b P(two members of the delegation are Labour and one is Conservative)

$$= 3 \times \frac{45}{100} \times \frac{44}{99} \times \frac{30}{98}$$

$$= \frac{9}{49}$$

> We need to find all the routes with 2 Ls and 1 C.

> There are three routes: $L_1L_2C_3$, $L_1C_2L_3$, $C_1L_2L_3$. Each fraction has 45, 44 and 30 on the top and 100, 99 and 98 on the bottom.

> In S1 you will not be expected to draw a full tree diagram as complicated as this but the **structure** of the tree diagram should help you calculate how many routes you need for a particular problem. In Book S2 the factorial notation will be introduced and a formula for calculating the number of arrangements in a more complicated situation will be given.

Example 18

Events A, B and C are defined in the sample space S such that $P(A) = 0.4$, $P(B) = 0.2$, $P(A \cap C) = 0.04$ and $P(B \cup C) = 0.44$. The events A and B are mutually exclusive and B and C are independent.

a Draw a Venn diagram to illustrate the relationship between the three events and the sample space

Find

b $P(B|C)$,   **c** $P(C)$,   **d** $P(B \cap C)$,   **e** $P(A' \cap B' \cap C')$,   **f** $P(C \cap B')$.

a

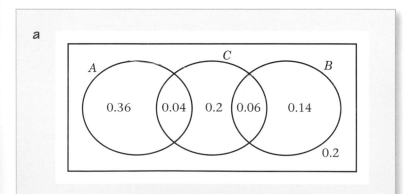

> A and B are mutually exclusive so they do not intersect. $P(A \cap C) = 0.04 > 0$ so A and C intersect. B and C are independent so they intersect as $P(B) > 0$ and $P(C) > 0$.

> The diagram without probabilities illustrates the relationship between events, but probabilities can be added to help with the rest of the question. Only the two black values can be added at this stage.

b $P(B|C)$ = as B and C are independent

$= 0.2$

c $P(B \cap C) = P(B) + P(C) - P(B \cup C)$

$P(B) \times P(C) = 0.2 + P(C) - 0.44$

$0.2\,P(C) = P(C) - 0.24$

$0.8\,P(C) = 0.24$

$P(C) = 0.3$

> Use the Addition Rule and replace $P(B \cap C)$ with $P(B) \times P(C)$ as B and C are independent. This gives a simple equation to be solved with $P(C)$ unknown.

d $P(B \cap C) = P(B) \times P(C)$

$= 0.2 \times 0.3$

$= 0.06$

> The red probabilities can now be added to the Venn diagram.

e $P(A' \cap B' \cap C') = 0.2$

> This is the region outside the three closed curves.

f $P(C \cap B') = 0.04 + 0.2 = 0.24$

> These values are taken from the Venn diagram.

Example 19

In a college there are 100 students taking A level French, German or Spanish. Of these students, 64 are female and the rest are male. There are 50 French students of whom 40 are female and 30 German students of whom 10 are female.

Find the probability that a randomly chosen student

a is taking Spanish,

b is male, given that the student is taking Spanish.

College records indicate that 70% of the French students, 80% of the German students and 60% of the Spanish students have applied for University.

A student is chosen at random.

c Find the probability that this student has applied for University.

d Given that the student had applied to University, find the probability that the student is studying French.

	Female	Male	**Totals**
French	40	10	50
German	10	20	30
Spanish	14	6	20
Totals	64	36	100

> Each row has a missing value that can be found by subtraction.

> This value is the 'Spanish' row and is found using the column totals

a $P(\text{Spanish}) = \dfrac{20}{100} = \dfrac{1}{5}$

b $P(\text{Male} \mid \text{Spanish}) = \dfrac{P(\text{Male and Spanish})}{P(\text{Spanish})}$

> Use the definition of conditional probability and substitute the values from the table.

$= \dfrac{\frac{6}{100}}{\frac{20}{100}} = \dfrac{6}{20} = \dfrac{3}{10}$

c $P(\text{Applied for Uni}) = \dfrac{50}{100} \times 0.70 + \dfrac{30}{100} \times 0.80 + \dfrac{20}{100} \times 0.60$

$= 0.71$

> Take each language in turn and find the number going to University by multiplying by the appropriate % then dividing by the total number of students i.e. 100.

d $P(\text{French} \mid \text{Applied to Uni}) = \dfrac{P(\text{French and Uni})}{P(\text{Uni})}$

$= \dfrac{\frac{50}{100} \times 0.70}{\frac{71}{100}}$

$= \dfrac{35}{71}$

> Use the definition of conditional probability and substitute the values from the table and **c**.

Mixed exercise 5G

1 The events A and B are such that $P(A) = \frac{1}{3}$, $P(B) = \frac{1}{4}$ and $P(A \cup B) = \frac{1}{2}$.

    a Show that A and B are independent.

    b Represent these probabilities in a Venn diagram.

    c Find $P(A \mid B')$.

2 A computer game has three levels and one of the objectives of every level is to collect a diamond. The probability of a randomly chosen player collecting a diamond on the first level is $\frac{4}{5}$, the second level is $\frac{2}{3}$ and the third level is $\frac{1}{2}$. The events are independent.

    a Draw a tree diagram to represent collecting diamonds on the three levels of the game.

    Find the probability that a randomly chosen player

    b collects all three diamonds,

    c collects only one diamond.

3 An online readers' club has 50 members. Glasses are worn by 15 members, 18 are left handed and 21 are female. There are four females who are left handed, three females who wear glasses and five members who wear glasses and are left handed. Only one member wears glasses, is left handed and female.

    a Draw a Venn diagram to represent these data.

    A member is selected at random. Find the probability that the member

    b is female, does not wear glasses and is not left handed,

    c is male, does not wear glasses and is not left handed,

    d wears glasses given that she is left handed and female.

4 For the events J and K,

$$P(J \cup K) = 0.5, \; P(J' \cap K) = 0.2, \; P(J \cap K') = 0.25.$$

a Draw a Venn diagram to represent the events J and K and the sample space S.

Find

b $P(J)$,   **c** $P(K)$,   **d** $P(J \mid K)$.

e Determine whether or not J and K are independent.

5 There are 15 coloured beads in a bag; seven beads are red, three are blue and five are green. Three beads are selected at random from the bag and not replaced. Find the probability that

a the first and second beads chosen are red and the third bead is blue or green,

b one red, one blue and one green bead are chosen.

6 A survey of a group of students revealed that 85% have a mobile phone, 60% have an MP3 player and 5% have neither phone nor MP3 player.

a Find the proportion of students who have both gadgets.

b Draw a Venn diagram to represent this information.

Given that a randomly selected student has a phone or an MP3 player,

c find the probability that the student has a mobile phone.

7 In a factory, machines A, B and C produce electronic components. Machine A produces 16% of the components, machine B produces 50% of the components and machine C produces the rest. Some of the components are defective. Machine A produces 4%, machine B 3% and machine C 7% defective components.

a Draw a tree diagram to represent this information.

Find the probability that a randomly selected component is

b produced by machine B and is defective,

c defective.

Given that a randomly selected component is defective,

d find the probability that it was produced by machine B.

8 A garage sells three types of fuel; U95, U98 and diesel. In a survey of 200 motorists buying fuel at the garage, 80 are female and the rest are male. Of the 90 motorists buying 'U95' fuel, 50 were female and of the 70 motorists buying diesel, 60 were male. A motorist does not buy more than one type of fuel.

Find the probability that a randomly chosen motorist

a buys U98 fuel,

b is male, given that the motorist buys U98 fuel.

Garage records indicate that 10% of the motorists buying U95 fuel, 30% of the motorists buying U98 fuel and 40% of the motorists buying diesel have their car serviced by the garage.

A motorist is chosen at random.

c Find the probability that this motorist has his or her car serviced by the garage.

d Given the motorist has his or her car serviced by the garage, find the probability that the motorist buys diesel fuel.

9 A study was made of a group of 150 children to determine which of three cartoons they watch on television. The following results were obtained:

> 35 watch Toontime
> 54 watch Porky
> 62 watch Skellingtons
> 9 watch Toontime and Porky
> 14 watch Porky and Skellingtons
> 12 watch Toontime and Skellingtons
> 4 watch Toontime, Porky and Skellingtons

a Draw a Venn diagram to represent these data.

Find the probability that a randomly selected child from the study watches

b none of the three cartoons,

c no more than one of the cartoons.

A child selected at random from the study only watches one of the cartoons.

d Find the probability that it was Skellingtons.

Two different children are selected at random from the study.

e Find the probability that they both watch Skellingtons.

10 The members of a wine tasting club are married couples. For any married couple in the club, the probability that the husband is retired is 0.7 and the probability that the wife is retired 0.4. Given that the wife is retired, the probability that the husband is retired is 0.8.

For a randomly chosen married couple who are members of the club, find the probability that

a both of them are retired,

b only one of them is retired,

c neither of them is retired.

Two married couples are chosen at random.

d Find the probability that only one of the two husbands and only one of the two wives is retired.

Summary of key points

1 P(event A or event B or both) = $P(A \cup B)$

P(both events A and B) = $P(A \cap B)$

P(not event A) = $P(A')$

2 Complementary probability

$P(A') = 1 - P(A)$

3 Addition rule

$P(A \cup B) = P(A) + P(B) - P(A \cap B)$

4 Conditional probability

$P(A$ given $B) = P(A|B) = \dfrac{P(A \cap B)}{P(B)}$

5 Multiplication rule

$P(A \cap B) = P(A|B) \times P(B)$ or $P(B|A) \times P(A)$

6 A and B are **independent** if

$P(A|B) = P(A)$ or $P(B|A) = P(B)$ or $P(A \cap B) = P(A) \times P(B)$

7 A and B are **mutually exclusive** if

$P(A \cap B) = 0$

Review Exercise

1 In a factory, machines *A*, *B* and *C* are all producing metal rods of the same length. Machine *A* produces 35% of the rods, machine *B* produces 25% and the rest are produced by machine *C*. Of their production of rods, machines *A*, *B* and *C* produce 3%, 6% and 5% defective rods respectively.

a Draw a tree diagram to represent this information.

b Find the probability that a randomly selected rod is

  i produced by machine *A* and is defective,

  ii is defective.

c Given that a randomly selected rod is defective, find the probability that it was produced by machine *C*.   **E**

2 Summarised opposite are the distances, to the nearest mile, travelled to work by a random sample of 120 commuters.

Distance (to the nearest mile)	Number of commuters
0–9	10
10–19	19
20–29	43
30–39	25
40–49	8
50–59	6
60–69	5
70–79	3
80–89	1

For this distribution,

a describe its shape,

b use linear interpolation to estimate its median.

The mid-point of each class was represented by x and its corresponding frequency by f giving

  $\sum fx = 3550$ and $\sum fx^2 = 138\,020$

c Estimate the mean and standard deviation of this distribution.

One coefficient of skewness is given by

$$\frac{3(\text{mean} - \text{median})}{\text{standard deviation}}$$

d Evaluate this coefficient for this distribution.

e State whether or not the value of your coefficient is consistent with your description in part **a**. Justify your answer.

f State, with a reason, whether you should use the mean or the median to represent the data in this distribution.

g State the circumstance under which it would not matter whether you used the mean or the median to represent a set of data.

3 A teacher recorded, to the nearest hour, the time spent watching television during a particular week by each child in a random sample. The times were summarised in a grouped frequency table and represented by a histogram.

One of the classes in the grouped frequency distribution was 20–29 and its associated frequency was 9. On the histogram the height of the rectangle representing that class was 3.6 cm and the width was 2 cm.

a Give a reason to support the use of a histogram to represent these data.

b Write down the underlying feature associated with each of the bars in a histogram.

c Show that on this histogram each child was represented by 0.8 cm^2.

The total area under the histogram was 24 cm^2.

d Find the total number of children in the group. *(E)*

4 a Give two reasons to justify the use of statistical models.

It has been suggested that there are seven stages involved in creating a statistical model. They are summarised below, with stages 3, 4 and 7 missing.

Stage 1. The recognition of a real-world problem.

Stage 2. A statistical model is devised.

Stage 3.

Stage 4.

Stage 5. Comparisons are made against the devised model.

Stage 6. Statistical concepts are used to test how well the model describes the real-world problem.

Stage 7.

b Write down the missing stages. *(E)*

5 The following table summarises the distances, to the nearest km, that 134 examiners travelled to attend a meeting in London.

Distance (km)	Number of examiners
41–45	4
46–50	19
51–60	53
61–70	37
71–90	15
91–150	6

a Give a reason to justify the use of a histogram to represent these data.

b Calculate the frequency densities needed to draw a histogram for these data.
(Do not draw the histogram.)

c Use interpolation to estimate the median Q_2, the lower quartile Q_1, and the upper quartile Q_3 of these data.

The mid-point of each class is represented by x and the corresponding frequency by f. Calculations then give the following values

$$\sum fx = 8379.5 \quad \text{and} \quad \sum fx^2 = 557\,489.75$$

d Calculate an estimate of the mean and an estimate of the standard deviation for these data.

One coefficient of skewness is given by

$$\frac{Q_3 - 2Q_2 + Q_1}{Q_3 - Q_1}.$$

e Evaluate this coefficient and comment on the skewness of these data.

f Give another justification of your comment in **e**.

(E)

6 Aeroplanes fly from City A to City B. Over a long period of time the number of minutes delay in take-off from City A was recorded. The minimum delay was five minutes and the maximum delay was 63 minutes. A quarter of all delays were at most 12 minutes, half were at most 17 minutes and 75% were at most 28 minutes. Only one of the delays was longer than 45 minutes.

An outlier is an observation that falls either 1.5 × (interquartile range) above the upper quartile or 1.5 × (interquartile range) below the lower quartile.

a On graph paper, draw a box plot to represent these data.

b Comment on the distribution of delays. Justify your answer.

c Suggest how the distribution might be interpreted by a passenger who frequently flies from City A to City B.

(E)

7 In a school there are 148 students in Years 12 and 13 studying Science, Humanities or Arts subjects. Of these students, 89 wear glasses and the others do not. There are 30 Science students of whom 18 wear glasses. The corresponding figures for the Humanities students are 68 and 44 respectively.

A student is chosen at random.

Find the probability that this student

a is studying Arts subjects,

b does not wear glasses, given that the student is studying Arts subjects.

Amongst the Science students, 80% are right-handed. Corresponding percentages for Humanities and Arts students are 75% and 70% respectively.

A student is again chosen at random.

c Find the probability that this student is right-handed.

d Given that this student is right-handed, find the probability that the student is studying Science subjects.

(E)

8 Over a period of time, the number of people x leaving a hotel each morning was recorded. These data are summarised in the stem and leaf diagram below.

Number leaving	3|2 means 32	Totals
2	7 9 9	(3)
3	2 2 3 5 6	(5)
4	0 1 4 8 9	(5)
5	2 3 3 6 6 6 8	(7)
6	0 1 4 5	(4)
7	2 3	(2)
8	1	(1)

For these data,

a write down the mode,

b find the values of the three quartiles.

Given that $\sum x = 1335$ and $\sum x^2 = 71\,801$, find

c the mean and the standard deviation of these data.

One measure of skewness is found using

$$\frac{3(\text{mean} - \text{mode})}{\text{standard deviation}}.$$

d Evaluate this measure to show that these data are negatively skewed.

e Give two other reasons why these data are negatively skewed. **E**

9 A bag contains nine blue balls and three red balls. A ball is selected at random from the bag and its colour is recorded. The ball is not replaced. A second ball is selected at random and its colour is recorded.

a Draw a tree diagram to represent the information.

Find the probability that

b the second ball selected is red,

c both balls selected are red, given that the second ball selected is red. **E**

10 For the events A and B,

$P(A \cap B') = 0.32$, $P(A' \cap B) = 0.11$ and $P(A \cup B) = 0.65$.

a Draw a Venn diagram to illustrate the complete sample space for the events A and B.

b Write down the value of $P(A)$ and the value of $P(B)$.

c Find $P(A|B')$.

d Determine whether or not A and B are independent. **E**

11 a Describe the main features and uses of a box plot.

Children from schools A and B took part in a fun run for charity. The times, to the nearest minute, taken by the children from school A are summarised in Figure 1.

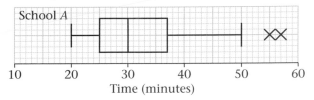

Figure 1

b i Write down the time by which 75% of the children in school A had completed the run.

ii State the name given to this value.

c Explain what you understand by the two crosses (×) on Figure 1.

For school B the least time taken by any of the children was 25 minutes and the longest time was 55 minutes. The three quartiles were 30, 37 and 50 respectively.

d On graph paper, draw a box plot to represent the data from school B.

e Compare and contrast these two box plots. **E**

12 Sunita and Shelley talk to each other once a week on the telephone. Over many weeks they recorded, to the nearest minute, the number of minutes spent in conversation on each occasion. The following table summarises their results.

Time (to the nearest minute)	Number of conversations
5–9	2
10–14	9
15–19	20
20–24	13
25–29	8
30–34	3

Two of the conversations were chosen at random.

a Find the probability that both of them were longer than 24.5 minutes.

The mid-point of each class was represented by x and its corresponding frequency by f, giving $\sum fx = 1060$.

b Calculate an estimate of the mean time spent on their conversations.

During the following 25 weeks they monitored their weekly conversations and found that at the end of the 80 weeks their overall mean length of conversation was 21 minutes.

c Find the mean time spent in conversation during these 25 weeks.

d Comment on these two mean values. **E**

13 A group of 100 people produced the following information relating to three attributes. The attributes were wearing glasses, being left-handed and having dark hair.

Glasses were worn by 36 people, 28 were left-handed and 36 had dark hair. There were 17 who wore glasses and were left-handed, 19 who wore glasses and had dark hair and 15 who were left-handed and had dark hair. Only 10 people wore glasses, were left-handed and had dark hair.

a Represent these data on a Venn diagram.

A person was selected at random from this group.

Find the probability that this person

b wore glasses but was not left-handed and did not have dark hair,

c did not wear glasses, was not left-handed and did not have dark hair,

d had only two of the attributes,

e wore glasses given they were left-handed and had dark hair. **E**

14 A survey of the reading habits of some students revealed that, on a regular basis, 25% read quality newspapers, 45% read tabloid newspapers and 40% do not read newspapers at all.

a Find the proportion of students who read both quality and tabloid newspapers.

b Draw a Venn diagram to represent this information.

A student is selected at random. Given that this student reads newspapers on a regular basis,

c find the probability that this student only reads quality newspapers. **E**

15 Figure 2 shows a histogram for the variable t which represents the time taken, in minutes, by a group of people to swim 500 m.

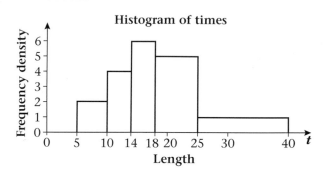

Histogram of times

Figure 2

a Copy and complete the frequency table for t.

t	5–10	10–14	14–18	18–25	25–40
Frequency	10	16	24		

b Estimate the number of people who took longer than 20 minutes to swim 500 m.

c Find an estimate of the mean time taken.

d Find an estimate for the standard deviation of *t*.

e Find an estimate for the median and quartiles for *t*.

One measure of skewness is found using

$$\frac{3(\text{mean} - \text{median})}{\text{standard deviation}}$$

f Evaluate this measure and describe the skewness of these data. **E**

16 A company assembles drills using components from two sources. Goodbuy supplies 85% of the components and Amart supplies the rest. It is known that 3% of the components supplied by Goodbuy are faulty and 6% of those supplied by Amart are faulty.

a Represent this information on a tree diagram.

An assembled drill is selected at random.

b Find the probability that it is not faulty. **E**

17 Data is coded using $y = \dfrac{x - 120}{5}$. The mean of the coded data is 24 and the standard deviation is 2.8. Find the mean and the standard deviation of the original data.

6

After completing this chapter you should be able to

- show diagrammatically, pairs of observations of variables
- be able to decide if there is a relationship between the variables
- put a numerical measure on the strength of this relationship
- simplify calculations by using coding
- calculate product moment correlation coefficients
- understand the limitations of product moment correlation coefficients.

You could use the above to see, for example, if there is any relationship between a person's pulse rate and the length of time they spent exercising, and, if there is, calculate a coefficient that is a measure of this relationship.

Correlation

If two variables are related to any extent, then changes in the value of one are related to changes in the value of the other. Such variables are examples of **bivariate data**. Bivariate data is data which comes in pairs e.g. (x, y) and there may or may not be a **relationship** between them.

6.1 **The association between two variables may be seen by plotting a scatter diagram.**

Example 1

10 students sat a Mathematics test and a Physics test. Both tests were marked out of 20.

Their marks are shown below.

Student	A	B	C	D	E	F	G	H	I	J
Mathematics mark	4	9	6	10	20	7	12	17	11	10
Physics mark	6	8	9	13	20	9	10	17	13	8

Draw a scatter diagram to represent these data.

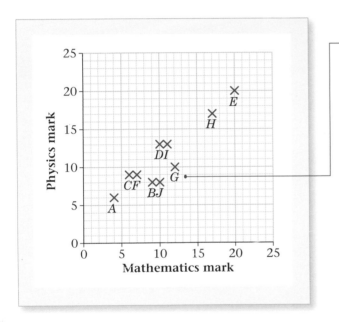

Point *G* is plotted at 12 horizontally and 10 vertically.

When drawing this scatter diagram you could have plotted Mathematics vertically and Physics horizontally.

From the scatter diagram you can see that the points lie approximately on a straight line. The higher the Mathematics mark the higher the Physics mark.

If, as in Example 1, both variables increase together they are said to be **positively correlated.**

If one variable increases as the other decreases they are said to be **negatively correlated**.

If no straight line (linear) pattern can be seen there is said to be **no correlation.** For example, there is no correlation between a person's height and how much they earn.

Typical scatter diagrams having positive, negative and no correlation are shown below.

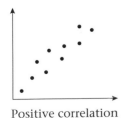

Positive correlation

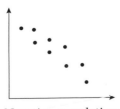

Negative correlation

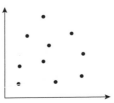

No correlation

Example 2

In the study of a city, the population density, in people/hectare, and the distance from the city centre, in km, was investigated by picking a number of sample areas with the following results.

Area	A	B	C	D	E	F	G	H	I	J
Distance (km)	0.6	3.8	2.4	3.0	2.0	1.5	1.8	3.4	4.0	0.9
Population density (people/hectare)	50	22	14	20	33	47	25	8	16	38

a Plot a scatter diagram.

b Describe the type of correlation.

c Interpret the correlation.

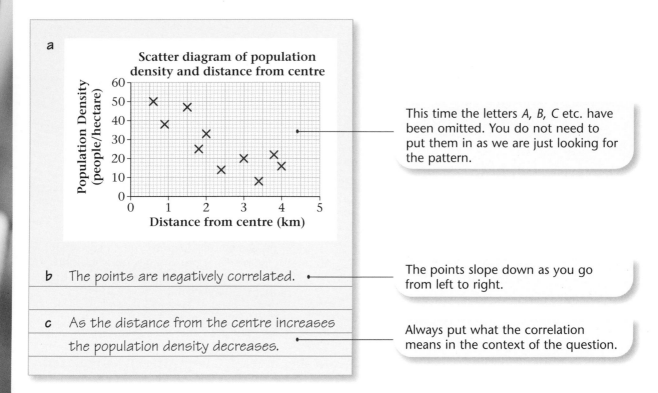

a

Scatter diagram of population density and distance from centre

This time the letters *A*, *B*, *C* etc. have been omitted. You do not need to put them in as we are just looking for the pattern.

b The points are negatively correlated.

The points slope down as you go from left to right.

c As the distance from the centre increases the population density decreases.

Always put what the correlation means in the context of the question.

If a horizontal line is drawn through the mean *y* value, $\bar{y}$, and a vertical line through the mean *x* value, $\bar{x}$, you can see the relationship between the two variables in another way.

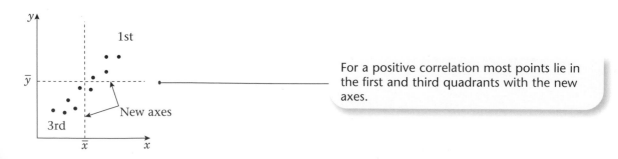

For a positive correlation most points lie in the first and third quadrants with the new axes.

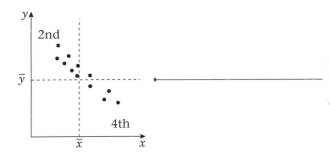

For a negative correlation most points lie in the second and fourth quadrants with the new axes.

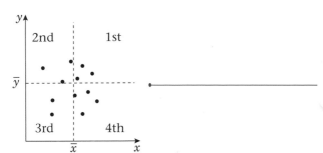

If there is no correlation the points are fairly distributed in all four quadrants with the new axes.

■ **For a positive correlation the points on the scatter diagram increase as you go from left to right. Most points lie in the first and third quadrants.**

■ **For a negative correlation the points on the scatter diagram decrease as you go from left to right. Most points lie in the second and fourth quadrants.**

■ **For no correlation the points lie fairly evenly in all four quadrants.**

Example 3

The scatter diagram shows the pulse rates, in beats per minute, and the breathing rates, in breaths per minute, of 12 people.

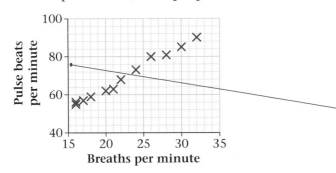

Notice the scales do not have to start at zero. The scales only have to cover the range of values.

Sometimes in the S1 exam you will be given scales and axes but if you have to choose pick sensible values.

Pick scales that go up by 1, 2, 5, 10, or by any two of these multiplied together, e.g. $2 \times 10 = 20$.

a Describe the type of correlation.

b Interpret the type of correlation.

a	Pulse rate and breaths per minute are positively correlated.
b	As breath rate increases so does the pulse rate.

Exercise 6A

1 The following scatter diagrams were drawn.

i   **ii**   **iii**

a Describe the type of correlation shown by each scatter diagram.

b Interpret each correlation.

2 Some research was done into the effectiveness of a weight reducing drug. Seven people recorded their weight loss and this was compared with the length of time for which they had been treated. A scatter diagram was drawn to represent these data.

a Describe the type of correlation shown by the scatter diagram.

b Interpret the correlation in context.

3 Eight metal ingots were chosen at random and measurements were made of their breaking strength (x) and their hardness (y). The results are shown in the table below.

x (tonne/cm)	5	7	7.4	6.8	5.4	7	6.6	6.4
y (hardness units)	50	70	85	70	75	60	65	60

a Draw a scatter diagram to represent these data.

b Describe and interpret the correlation between the variables 'hardness' and 'breaking strength'.

4 For each of the following data sets plot a scatter diagram, and then describe the correlation.

a

x	1	2.4	3.6	2.2	4.3	3.3	4.0	0.6
y	6.0	9.0	15.8	7.1	18.6	12.1	15.0	3.7

b

x	123	160	285	210	150	240	280	115	180
y	75	70	50	65	70	55	50	80	70

5 The table shows the armspan, in cm, and the height, in cm, of 10 adult men.

Height x (cm)	155	160	173	192	181	178	199	166	158	173
Armspan y (cm)	147	159	165	180	170	173	186	162	153	168

 a Draw a scatter diagram to represent these data.

 b Describe and interpret the correlation between the two variables 'height' and 'armspan'.

6 Eight students were asked to estimate the mass of a bag of sweets in grams. First they were asked to estimate the mass without touching the bag and then they were told to pick the bag up and estimate the mass again. The results are shown in the table below.

Student	A	B	C	D	E	F	G	H
Estimate of mass not touching bag (g)	25	18	32	27	21	35	28	30
Estimate of mass holding bag (g)	16	11	20	17	15	26	22	20

 a Draw a scatter diagram to represent these data.

 b Describe and interpret the correlation between the two variables.

6.2 You can calculate measures for the variability of bivariate data.

In Chapter 3 you found the variance of a set of data by using the formula

Variance $= \dfrac{\sum(x - \bar{x})^2}{n}$.

In correlation we write $\sum(x - \bar{x})^2 = \sum(x - \bar{x})(x - \bar{x})$ as S_{xx}

The S stands for Sum

■ $S_{xx} = \sum(x - \bar{x})^2$

So variance $= \dfrac{S_{xx}}{n}$

In a similar way

■ $S_{yy} = \sum(y - \bar{y})^2$

You can also define a quantity known as the **co-variance**.

Co-variance $= \dfrac{\sum(x - \bar{x})(y - \bar{y})}{n}$ and again we use the shorthand

■ $S_{xy} = \sum(x - \bar{x})(y - \bar{y})$

Example 4

The head circumference in cm (x) and gestation period in weeks (y) for new-born babies at a certain clinic over a period of time were as follows.

Baby	A	B	C	D	E	F
Head circumference (x)	31.1	33.3	30.0	31.5	35.0	30.2
Gestation period (y)	36	37	38	38	40	40

Find S_{xx} for these data.

$$\sum(x - \bar{x})^2 = [(31.85 - 31.1)^2 + (33.3 - 31.85)^2 + (31.85 - 30.0)^2$$
$$+ (31.85 - 31.5)^2 + (35.0 - 31.85)^2 + (31.85 - 30.2)^2]$$
$$= 0.5625 + 2.1025 + 3.4225 + 0.1225 + 9.9225 + 2.7225$$
$$= 18.855$$

There is an easier way of calculating S_{xx}, S_{yy} and S_{xy}

In Chapter 3 an easier way to calculate variance was found by using the formulae

variance $= \dfrac{\sum x^2}{n} - \bar{x}^2$.

Since variance $= \dfrac{S_{xx}}{n}$ you can see that

$S_{xx} = n \times \text{variance} = n \times \left\{ \dfrac{\sum x^2}{n} - \bar{x}^2 \right\}$ ⟵ Using the easy formulae for variance.

$\qquad\qquad = \sum x^2 - n\bar{x}^2$

$\qquad\qquad = \sum x^2 - \dfrac{n(\sum x)^2}{n^2}$ ⟵ Since $\bar{x} = \dfrac{\sum x}{n}$

$\qquad\qquad = \sum x^2 - \dfrac{(\sum x)^2}{n}$

■ $S_{xx} = \sum x^2 - \dfrac{(\sum x)^2}{n}$ ⟵ This formula is given in the formula booklet.

If you treat S_{yy} in a similar way you get:

■ $S_{yy} = \sum y^2 - \dfrac{(\sum y)^2}{n}$ ⟵ This formula is given in the formula booklet.

You can also show that:

■ $S_{xy} = \sum xy - \dfrac{\sum x \sum y}{n}$ ⟵ Again this formula is given in the formula booklet.

You do not have to be able to derive these formulae but you need to be able to use them to work out a measure of correlation between two variables.

Example 5

The head circumference in cm (x) and gestation period in weeks (y) for new-born babies at a certain clinic over a period of time were as follows.

Baby	A	B	C	D	E	F
Head circumference (x)	31.1	33.3	30.0	31.5	35.0	30.2
Gestation period (y)	36	37	38	38	40	40

Find S_{xy} for these data.

$$S_{xy} = \sum xy - \frac{\sum x \sum y}{n}$$

$$\sum xy = 1119.6 + 1232.1 + 1140 + 1197 + 1400 + 1208 = 7296.7$$

$$\sum x = 191.1 \qquad \sum y = 229$$

$$S_{xy} = 7296.7 - \frac{191.1 \times 229}{6} = 7296.7 - 7293.65 = 3.05$$

6.3 **You can get a measure of the amount of correlation between two variables by using the product moment correlation coefficient r.**

This is defined as

■ $r = \dfrac{S_{xy}}{\sqrt{S_{xx}S_{yy}}}$

If you use a statistical calculator it may give you the value of the product moment correlation coefficient directly. To just write down the answer in an examination is risky, since no marks will be given if it is incorrect. It is also possible that you may be asked to find intermediate values in the calculation such as S_{xx}, S_{yy}, etc., as well as the value of r.

Example 6

The number of vehicles, x millions, and the number of accidents y thousands, in 15 different countries were:

Country	A	B	C	D	E	F	G	H	I	J	K	L	M	N	O
Vehicles (x) millions	8.6	13.4	12.8	9.3	1.3	9.4	13.1	4.9	13.5	9.6	7.5	9.8	23.3	21	19.4
Accidents (y) thousands	33	51	30	48	12	23	46	18	36	50	34	35	95	99	69

Calculate the product moment correlation coefficient for the number of vehicles and the number of accidents.

Using a calculator we find

$\sum x = 176.9$

$\sum y = 679$

$\sum x^2 = 2576.47$

$\sum y^2 = 39\,771$

$\sum xy = 9915.3$

Some of these may be given in the question.

$S_{xx} = \sum x^2 - \dfrac{(\sum x)^2}{n}$

Standard formula given in formula booklet.

$\qquad = 2576.47 - \dfrac{176.9^2}{15} = 490.23$

$S_{yy} = \sum y^2 - \dfrac{(\sum y)^2}{n}$

$\qquad = 39\,771 - \dfrac{679^2}{15} = 9034.93$

$S_{xy} = \sum xy - \dfrac{\sum x \sum y}{n}$

$\qquad = 9915.3 - \dfrac{176.9 \times 679}{15} = 1907.63$

$r = \dfrac{S_{xy}}{\sqrt{S_{xx}S_{yy}}}$

$\quad = \dfrac{1907.63}{\sqrt{490.23 \times 9034.93}} = 0.906\ldots$

The value of the correlation coefficient is 0.906. This is a positive correlation.

The greater the number of vehicles the higher the number of accidents.

Exercise 6B

1 Given $\sum x = 18.5$ $\quad \sum x^2 = 36$ $\quad n = 10$ find the value of S_{xx}.

2 Given $\sum y = 25.7$ $\quad \sum y^2 = 140$ $\quad n = 5$ find the value of S_{yy}.

3 Given $\sum x = 15$ $\quad \sum y = 35$ $\quad \sum xy = 91$ $\quad n = 5$ find the value of S_{xy}.

4 Given that $S_{xx} = 92$, $S_{yy} = 112$ and $S_{xy} = 100$ find the value of the product moment correlation coefficient.

5 Given the following summary data,

$$\sum x = 367 \quad \sum y = 270 \quad \sum x^2 = 33845 \quad \sum y^2 = 12976 \quad \sum xy = 17135 \quad n = 6$$

calculate the product moment correlation coefficient (r) using the formula:

$$r = \frac{S_{xy}}{\sqrt{S_{xx}S_{yy}}}$$

6 The ages, a years, and heights, h cm, of seven members of a team were recorded. The data were summarised as follows:

$$\sum a = 115 \quad \sum a^2 = 1899 \quad S_{hh} = 571.4 \quad S_{ah} = 72.1$$

a Find S_{aa}.

b Find the value of the product moment correlation coefficient between a and h.

c Describe and interpret the correlation between the age and height of these seven people based on these data.

7 In research on the quality of bacon produced by different breeds of pig, data were obtained about the leanness (l) and taste (t) of the bacon. The data is shown in the table.

Leanness l	1.5	2.6	3.4	5.0	6.1	8.2
Taste t	5.5	5.0	7.7	9.0	10.0	10.2

a Find S_{ll}, S_{tt} and S_{lt}.

b Calculate the product moment correlation coefficient between l and t using the values found in **a**. If you have a calculator that will work out r use it to check your answer.

8 Eight children had their IQ measured and then took a general knowledge test. Their IQ, (x), and their marks, (y), for the test were summarised as follows:

$$\sum x = 973 \quad \sum x^2 = 120123 \quad \sum y = 490 \quad \sum y^2 = 33000 \quad \sum xy = 61595.$$

a Calculate the product moment correlation coefficient.

b Describe and interpret the correlation coefficient between IQ and general knowledge.

9 In a training scheme for young people, the average time taken for each age group to reach a certain level of proficiency was measured. The data are shown in the table.

Age x (years)	16	17	18	19	20	21	22	23	24	25
Average time y (hours)	12	11	10	9	11	8	9	7	6	8

a Find S_{xx}, S_{yy} and S_{xy}.

b Use your answers to calculate the product moment correlation coefficient (r).

c Describe and interpret the relationship between average time and age.

6.4 You can determine the strength of the linear relationship between the variables by looking at the value of the product moment correlation.

The value of r varies between -1 and 1.

If $r = 1$ there is a perfect positive linear correlation between the two variables (all points fit a straight line with positive gradient).

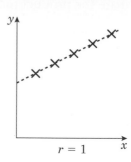

$r = 1$

If $r = -1$ there is a perfect negative linear correlation between the two variables (all points fit a straight line with negative gradient).

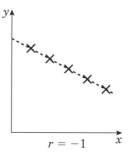

$r = -1$

If r is zero (or close to zero) there is no linear correlation; this does not, however, exclude any other sort of relationship.

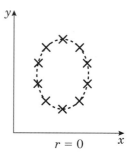

$r = 0$

Values of r between 1 and 0 indicate a greater or lesser degree of positive correlation. The closer to 1 the better the correlation, the closer to 0 the worse the correlation.

Values of r between -1 and 0 indicate a greater or lesser degree of negative correlation. The closer to -1 the better the correlation, the closer to 0 the worse the correlation.

Example 7

The scatter diagrams show various degrees of correlation.

a   **b**    **c**    **d**

Match the diagrams with the product moment correlation coefficients below.

$r = -0.31$ $r = -0.94$ $r = 0.55$ $r = 0.97$.

a = 0.97	**a** is a positive correlation that is close to 1.
b = −0.94	**b** is a negative correlation that is close to −1.
c = 0.55	**c** is a positive correlation that is not very close to 1 or 0.
d = −0.31	**d** is a negative correlation that is closer to 0 than to 1.

6.5 Understanding the limitations of the product moment correlation coefficient.

Even if two variables are associated and have a linear correlation, it does not necessarily mean that a change in one of the variables causes a change in the other variable.

Example 8

The number of cars on the road has increased, and the number of DVD recorders bought has decreased. Is there a correlation between these two variables?

Buying a car does not necessarily mean
that you will not buy a DVD recorder, so
we cannot say there is a correlation
between the two.

Variables are often linked only through a third variable. This can be shown in changes that take place over time.

Example 9

Over the past 10 years the memory capacity of personal computers has increased, and so has the average life expectancy of people in the western world. Is there a correlation between these two variables?

The two are not connected, but both
are due to scientific development over
time.

Exercise 6C

1 The following product moment correlation coefficients were calculated

i −0.96   **ii** −0.35   **iii** 0   **iv** 0.72

Write down the coefficient that

a shows the least correlation,   **b** shows the most correlation.

2 Here are some product moment correlation coefficients.

i −1,   **ii** −0.5,   **iii** 0,   **iv** 0.5,   **v** 1.

Write down which one shows

a perfect negative correlation,   **b** zero correlation.

3 Ahmed works out the product moment correlation coefficient between the heights of a group of fathers and the heights of their sons to be 0.954. Write down what this tells you about the relationship between their heights.

4 Maria draws some scatter diagrams. They are shown below.

a    **b**

c    **d**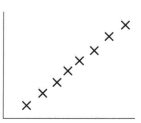

Write down which scatter diagram shows:

i a correlation of +1,

ii a correlation that could be described as strong positive correlation,

iii a correlation of about −0.97,

iv a correlation that shows almost no correlation.

5 Jake finds that the product moment correlation coefficients between two variables x and y is 0.95. The product moment correlation coefficient between two other variables s and t was −0.95. Discuss how these two coefficients differ.

6 Patsy collects some data to find out if there is any relationship between the numbers of car accidents and computer ownership. She calculates the product moment correlation coefficient between the two variables. There is a strong positive correlation. She says as car accidents increase so does computer ownership. Write down whether or not this is sensible. Give reasons for your answer.

7 Raj collects some data to find out whether there is any relationship between the height of students in his year group and the pass rate in driving tests. He finds that there is a strong positive correlation. He says that as height increases, so does your chance of passing your driving test. Is this sensible? Give reasons for your answer.

6.6 **Simplifying the calculation of the product moment correlation coefficient by using a method of coding, when the numbers are large.**

You can subtract any number from the x values, since this only moves the axis. You can divide the result by any number, since this only changes the scale.

The correlation coefficient is unaffected by either of these operations. The number you subtract and divide by can be selected in any way you choose, but it is common sense to make their values such that the resulting numbers are small.

■ **You can rewrite the variables x and y by using the coding**

$$p = \frac{x - a}{b} \text{ and } q = \frac{y - c}{d}$$

where a, b, c and d are suitable numbers to be chosen.

■ **The product moment correlation r is not affected by coding.**

Example 10

Calculate the product moment correlation coefficient for x and y if the values of x and y are as shown in the table below.

x	1020	1032	1028	1034	1023	1038
y	320	335	345	355	360	380

You may use the coding:

$$p = \frac{x - 1020}{1} \qquad q = \frac{y - 300}{5}.$$

p	q	p^2	q^2	pq
0	4	0	16	0
12	7	144	49	84
8	9	64	81	72
14	11	196	121	154
3	12	9	144	36
18	16	324	256	288
$\sum p = 55$	$\sum q = 59$	$\sum p^2 = 737$	$\sum q^2 = 667$	$\sum pq = 634$

Here $a = 1020$ and $b = 1$. These were chosen because 1020 was the smallest x value and the numbers then left had no common divisor.

In the same way $c = 300$ and $d = 5$ since the numbers that are left have 5 as a common divisor.

(We could have used 320 and made the values even smaller, but wished to show that *any* number will do.)

$$S_{pp} = 737 - \frac{55^2}{6} = 232.83$$

$$S_{qq} = 667 - \frac{59^2}{6} = 86.83$$

$$S_{pq} = 634 - \frac{55 \times 59}{6} = 93.17$$

$$r = \frac{S_{pq}}{\sqrt{S_{pp}S_{qq}}} = \frac{93.17}{\sqrt{232.83 \times 86.83}}$$

$$= 0.655$$

Remember the product moment correlation coefficient is not affected by coding.

Exercise 6D

1 Coding is to be used to work out the value of the product moment correlation coefficient for the following sets of data. Suggest a suitable coding for each.

a

x	2000	2010	2015	2005	2003	2006
y	3	6	21	6	9	18

b

s	100	300	200	400	300	700
t	2	0	1	3	3	6

2 For the two variables x and y the coding of $A = x - 7$ and $B = y - 100$ is to be used.
The product moment correlation coefficient for A and B is found to be 0.973.
What is the product moment correlation coefficient for x and y?

3 Use the coding: $p = x$ and $q = y - 100$ to work out the product moment correlation coefficient for the following data.

x	0	5	3	2	1
y	100	117	112	110	106

4 The product moment correlation is to be worked out for the following data set using coding.

x	50	40	55	45	60
y	4	3	5	4	6

a Using the coding $p = \frac{x}{5}$ and $t = y$ find the values of S_{pp}, S_{tt} and S_{pt}.
b Calculate the product moment correlation between p and t.
c Write down the product moment correlation between x and y.

5 The tail length (*t* cm) and the mass (*m* grams) for each of eight woodmice were measured. The data is shown in the table.

t (cm)	8.5	7.5	8.6	7.3	8.1	7.5	8.0	7.8
m (g)	28	22	26	21	25	20	20	22

a Using the coding $x = t - 7.3$ and $y = m - 20$ complete the following table

x	1.2			0				0.5
y	8			1				

b Find S_{xx}, S_{yy} and S_{xy}.

c Calculate the value of the product moment correlation coefficient between *x* and *y*.

d Write down the product moment correlation coefficient between *t* and *m*.

e Write down the conclusion that can be drawn about the relationship between tail length and mass of woodmice.

6 A shopkeeper thinks that the more newspapers he sells in a week the more sweets he sells. He records the amount of money (*m* pounds) that he takes in newspaper sales and also the amount of money he takes in sweet sales (*s* pounds) each week for seven weeks. The data are shown are the following table.

Newspaper sales (*m* pounds)	380	402	370	365	410	392	385
Sweet sales (*s* pounds)	560	543	564	573	550	544	530

a Use the coding $x = m - 365$ and $y = s - 530$ to find S_{XX}, S_{YY} and S_{XY}.

b Calculate the product moment correlation coefficient for *m* and *s*.

c State, with a reason, whether or not what the shopkeeper thinks is correct.

Mixed exercise 6E

1 The following table shows the distance (*x*) in miles and the cost (*y*) in pounds of each of 10 taxi journeys.

x (miles)	8	6.5	4	2.5	5.5	9	2	10	4.5	7.5
y (pounds)	10.2	8.8	7.2	5.7	7.4	11.0	5.2	12.0	6.4	10.0

a Draw a scatter diagram to represent these data.

b Describe and interpret the correlation between the two variables.

2 The following scatter diagrams were drawn.

i Length and age of snakes — Length (vertical axis), Age (horizontal axis)

ii Population — % Drop in wages (vertical axis), % Unemployment (horizontal axis)

iii Men — Height (vertical axis), Age (horizontal axis)

    a State whether each shows positive, negative or no correlation.

    b Interpret each scatter diagram in context.

3 The following scatter diagrams were drawn by a student.

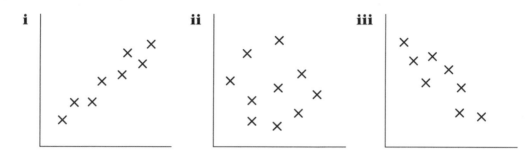

i   **ii**   **iii**

The student calculated the product moment correlation coefficient for each set of data.

The values were:

    a −0.12   **b** 0.87   **c** −0.81

Write down which value corresponds to each scatter diagram. Give a reason for your answer.

4 The product moment correlation coefficient for a person's age and his score on a memory test is −0.86. Interpret this value.

5 Wai wants to know whether the 10 people in her group are as good at Science as they are at Art. She collected the end of term test marks for Science (s), and Art (a), and coded them using $x = \dfrac{s}{10}$ and $y = \dfrac{a}{10}$.

The data she collected can be summarised as follows,

$$\sum x = 67 \qquad \sum x^2 = 465 \qquad \sum y = 65 \qquad \sum y^2 = 429 \qquad \sum xy = 434.$$

    a Work out the product moment correlation coefficient for x and y.

    b Write down the product moment correlation coefficient for s and a.

    c Write down whether or not it is it true to say that the people in Wai's group who are good at Science are also good at Art. Give a reason for your answer.

6 Nimer thinks that oranges that are very juicy cost more than those that are not very juicy. He buys 20 oranges from different places, and measures the amount of juice (j ml), that each orange produces. He also notes the price (p) of each orange.

The data can be summarised as follows,

$$\sum j = 979 \qquad \sum p = 735 \qquad \sum j^2 = 52\,335 \qquad \sum p^2 = 32\,156 \qquad \sum jp = 39\,950.$$

a Find S_{jj}, S_{pp} and S_{jp}.

b Using your answers to **a** calculate the product moment correlation coefficient.

c Describe the type of correlation between the amount of juice and the cost and state, with a reason, whether or not Nimer is correct.

7 The following table shows the values of two variables v and m.

v	50	70	60	82	45	35	110	70	35	30
m	140	200	180	210	120	100	200	180	120	60

The results were coded using $x = v - 30$ and $y = \dfrac{m}{20}$.

a Complete the table for x and y.

x	20	40			15			40		0
y	7	10		10.5	6					3

b Calculate S_{xx}, S_{yy} and S_{xy}.
(You may use $\sum x = 287$, $\sum x^2 = 13\,879$, $\sum y = 75.5$, $\sum y^2 = 627.25$, $\sum xy = 2661$.)

c Using your answers to **b** calculate the product moment correlation coefficient for x and y.

d Write down the product moment correlation coefficient for v and m.

e Describe and interpret your product moment correlation coefficient for v and m.

8 Each of 10 cows was given an additive (x) every day for four weeks to see if it would improve their milk yield (y). At the beginning the average milk yield per day was 4 gallons. The milk yield of each cow was measured on the last day of the four weeks. The data collected is shown in the table.

Cow	A	B	C	D	E	F	G	H	I	J
Additive, x (25 gm units)	1	2	3	4	5	6	7	8	9	10
Yield, y (gallons)	4.0	4.2	4.3	4.5	4.5	4.7	5.2	5.2	5.1	5.1

a Draw a scatter diagram of these data.

b Write down any conclusions you can draw from the scatter diagram.

c From the diagram write down, with a reason, the amount of additive that could be given to each cow to maximise yield and minimise cost.

d The product moment correlation coefficient is to be calculated for the first seven cows. Write down why you think cows H, I and J are being left out for this calculation.

e Use the values $S_{xx} = 28$, $S_{yy} = 0.90857$ and $S_{xy} = 4.8$ to calculate the product moment correlation coefficient for the seven cows.

f Write down, with a reason, how the product moment correlation coefficient for all 10 cows would differ from your answer to **e.**

9 The following table shows the engine size (c), in cubic centimetres, and the fuel consumption (f), in miles per gallon to the nearest mile, for 10 car models.

c (cm)^3	1000	1200	1400	1500	1600	1800	2000	2200	2500	3000
f (mpg)	46	42	43	39	41	37	35	29	28	25

a On graph paper draw a scatter diagram to represent these data.

b Write down whether the correlation coefficient between c and f is positive or negative. Give a reason for your answer.

The data can be summarised as follows:

$\sum cf = 626\ 100$, $\sum c = 18\ 200$, $\sum f = 365$

c Calculate S_{cf}

d The product moment correlation coefficient could be found by using coding. Suggest suitable coding.

10 In a study on health, a clinic measured the age, (a years), and the diastolic blood pressure, (d in mm of mercury), of eight patients. The table shows the results.

a (years)	20	35	50	25	60	45	25	70
d (mm)	55	60	80	85	75	85	70	85

a Using the coding $x = \dfrac{a}{5}$ and $y = \dfrac{d}{5} - 11$ calculate S_{xx}, S_{yy} and S_{xy}.

b Using your answers to **a** work out the product moment correlation coefficient for x and y.

c Write down the product moment correlation coefficient between a and d.

d Interpret your answer to **c.**

Summary of key points

1 For a positive correlation the points on the scatter diagram increase as you go from left to right. Most points lie in the first and third quadrants.

For a negative correlation the points on the scatter diagram decrease as you go from left to right. Most points lie in the second and fourth quadrants.

For no correlation, the points lie in all four quadrants.

2 $S_{xx} = \sum(x - \bar{x})^2 = \sum x^2 - \dfrac{(\sum x)^2}{n}$

$S_{yy} = \sum(y - \bar{y})^2 = \sum y^2 - \dfrac{(\sum y)^2}{n}$

$S_{xy} = \sum(x - \bar{x})(y - \bar{y}) = \sum xy - \dfrac{\sum x \sum y}{n}$

3 $r = \dfrac{S_{xy}}{\sqrt{S_{xx}S_{yy}}}$

4 r is a measure of linear relationship

$r = 1 \Rightarrow$ perfect positive linear correlation

$r = -1 \Rightarrow$ perfect negative correlation

$r = 0 \Rightarrow$ no linear correlation

5 You can rewrite the variables x and y by using the coding $p = \dfrac{x - a}{b}$ and $q = \dfrac{y - c}{d}$.

r is not affected by coding.

7

Regression

After completing this chapter you should be able to

- understand the idea of **independent** and **dependent variables**,
- work out the **regression equation of a line** which best fits the trend of the points on a scatter diagram,
- apply and interpret the regression equation.

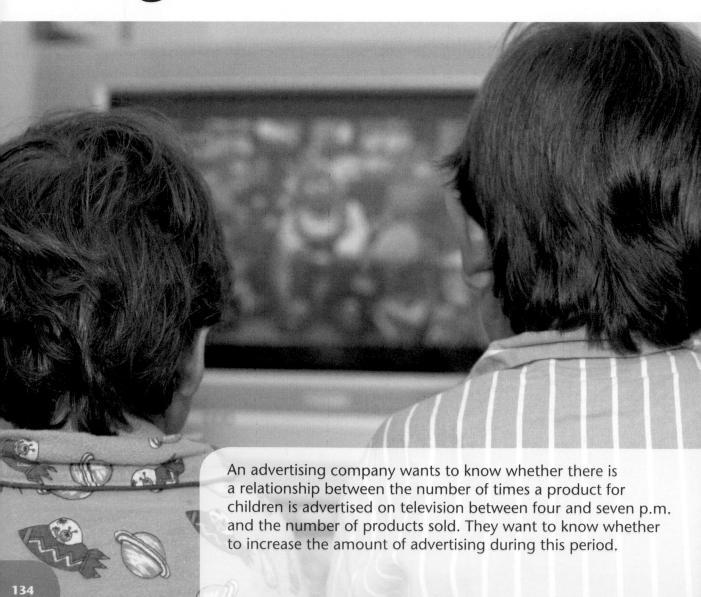

An advertising company wants to know whether there is a relationship between the number of times a product for children is advertised on television between four and seven p.m. and the number of products sold. They want to know whether to increase the amount of advertising during this period.

7.1 The rule $y = a + bx$ connecting the variables x and y allows the value of y to be predicted for any given value of x.

- $y = a + bx$ is the equation of a straight line

- If $y = a + bx$ then a (sometimes called the **intercept**) is where the line cuts the y-axis (the line $x = 0$) and b is the amount by which y increases for an increase of 1 in x, (b is called the **gradient** of the line).

Example 1

For the line $y = 2x + 3$

a write down the gradient and intercept on the y-axis of the line,

b sketch the line.

a The gradient = 2 and the intercept on the y-axis = 3

Compare $y = 2x + 3$ with $y = a + bx$. From this $b = 2$ and $a = 3$.

b

To draw the line start at (0, 3) and then, using the gradient, move along one unit and up two units. This gives a second point (1, 5) and hence a line can be drawn.

For each increase of one in x, y increases by two.

If the points on a **scatter diagram** follow a **linear pattern** a straight line can be used as a model for the relationship.

The **line of best fit** is the straight line that best models the relationship.

Example 2

A company would like to predict what the sales of their new shops are going to be. The yearly sales of nine of their existing shops are known and the sizes of the population of the towns in which each shop is sited are found. The size of the towns, (x) in1000s, and the sales, (y) in £1000s, are given in the table.

Town size x 1000	18	26	28	36	48	52	54	60	50
Sales y (£1000)	48	62	53	65	72	70	73	76	50

Draw the scatter diagram for these data, and draw by eye the line of best fit through the points.

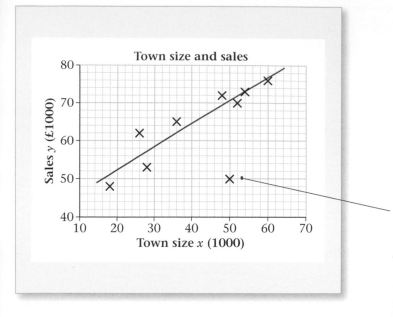

The line is drawn so that points lie fairly evenly either side of it and in the direction of the crosses.

This process is not very accurate and you will see how a formula can be used to calculate the equation of the line of best fit.

This point is outside the trend and is ignored. (Perhaps someone copied a figure incorrectly.)

7.2 Independent and dependent variables.

- An **independent (or explanatory) variable** is one that is set independently of the other variable. It is plotted along the x-axis.

- A **dependent (or response) variable** is one whose values are determined by the values of the independent variable. It is plotted along the y-axis.

Example 3

A company wants to predict sales of a new album by a pop group. They know the yearly sales of a number of existing albums by the same group and the number of stores that stock each album. Which is the independent variable and which is the dependent variable?

The yearly sales of each album depend on the number of stores that stock it.

Therefore the independent variable is the number of stores and the dependent variable is the sales.

7.3 The values for a and b that make the sum of the squares of the residuals a minimum can be calculated using the formulae $b = \dfrac{S_{xy}}{S_{xx}}$ and $a = \bar{y} - b\bar{x}$.

■ For each point on a scatter diagram you can express y in terms of x as $y = (a + bx) + e$, where e is the vertical distance from the line of best fit. This is shown in the diagram.

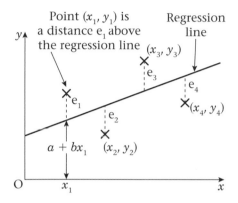

The values e_1, e_2, e_3 etc are known as **residuals**.

■ The line that minimises the sum of the squares of the residuals is called the least squares regression line. That is to say $\sum e^2$ is a minimum. The line is called the **regression line of y on x**.

■ The **equation of the regression line** of y on x is: $y = a + bx$ where

$$b = \frac{S_{xy}}{S_{xx}} \text{ and } a = \bar{y} - b\bar{x}$$

S_{xy} and S_{xx} are the values used in the previous chapter. They can be found in the formula book.

Example 4

The results from an experiment in which different masses were placed on a spring and the resulting length of the spring measured, are shown below.

Mass, x (kg)	20	40	60	80	100
Length, y (cm)	48	55.1	56.3	61.2	68

a Calculate S_{xx} and S_{xy}.

(You may use $\sum x = 300$, $\sum x^2 = 22\,000$, $\bar{x} = 60$, $\sum xy = 18\,238$, $\sum y^2 = 16\,879.14$, $\sum y = 288.6$, $\bar{y} = 57.72$)

b Calculate the regression line of y on x.

a $S_{xx} = \sum x^2 - \dfrac{(\sum x)^2}{n}$

$\qquad = 22\,000 - \dfrac{300^2}{5}$

$\qquad = 4000$

$S_{xy} = \sum xy - \dfrac{\sum x \sum y}{n}$

$\qquad = 18\,238 - \dfrac{300 \times 288.6}{5}$

$\qquad = 922$

b $b = \dfrac{S_{xy}}{S_{xx}} = \dfrac{922}{4000} = 0.2305$

$a = \bar{y} - b\bar{x}$

$\qquad = 57.72 - 0.2305 \times 60$

$\qquad = 43.89$

$y = 43.89 + 0.2305x$

> Here you apply the standard formula for S_{xx} and S_{xy}. These were described in Chapter 6 and are also given in the formula booklet.

> Again these standard formula for a and b are given in the formula booklet.

> Remember to write the equation at the end. The numbers should be given to at least three significant figures.

Exercise 7A

1 An NHS trust has the following data on operations.

Number of operating theatres	5	6	7	8
Number of operations carried out per day	25	30	35	40

Which is the independent and which is the dependent variable?

2 A park ranger collects data on the number of species of bats in a particular area.

Number of suitable habitats	10	24	28
Number of species	1	2	3

Which is the independent and which is the dependent variable?

3 The equation of a regression line in the form $y = a + bx$ is to be found. Given that $S_{xx} = 15$, $S_{xy} = 90, \bar{x} = 3$ and $\bar{y} = 15$ work out the values of a and b.

4 Given $S_{xx} = 30$, $S_{xy} = 165, \bar{x} = 4$ and $\bar{y} = 8$ find the equation of the regression line of y on x.

5 The equation of a regression line is to be found. The following summary data is given.

$\qquad S_{xx} = 40, \qquad S_{xy} = 80, \qquad \bar{x} = 6, \qquad \bar{y} = 12.$

Find the equation of the regression line in the form $y = a + bx$.

6 Data is collected and summarised as follows.

$$\Sigma x = 10 \qquad \Sigma x^2 = 30 \qquad \Sigma y = 48 \qquad \Sigma xy = 140 \qquad n = 4.$$

a Work out $\bar{x}, \bar{y}, S_{xx}$ and S_{xy}.

b Find the equation of the regression line of y on x in the form $y = a + bx$.

7 For the data in the table:

x	2	4	5	8	10
y	3	7	8	13	17

a calculate S_{xx} and S_{xy},

b find the equation of the regression line of y on x in the form $y = a + bx$.

8 A field was divided into 12 plots of equal area. Each plot was fertilised with a different amount of fertilizer (h). The yield of grain (g) was measured for each plot. Find the equation of the regression line of g on h in the form $g = a + bh$ given the following summary data.

$$\Sigma h = 22.09 \qquad \Sigma g = 49.7 \qquad \Sigma h^2 = 45.04 \qquad \Sigma g^2 = 244.83 \qquad \Sigma hg = 97.778 \qquad n = 12$$

9 An accountant monitors the number of items produced per month by a company (n) together with the total production costs (p). The table shows these data.

Number of items, n, (1000s)	21	39	48	24	72	75	15	35	62	81	12	56
Production costs, p, (£1000s)	40	58	67	45	89	96	37	53	83	102	35	75

(You may use $\Sigma n = 540 \qquad \Sigma n^2 = 30\,786 \qquad \Sigma p = 780 \qquad \Sigma p^2 = 56\,936 \qquad \Sigma np = 41\,444$)

a Calculate S_{nn} and S_{np}.

b Find the equation of the regression line of p on n in the form $p = a + bn$.

10 The relationship between the number of coats of paint applied to a boat and the resulting weather resistance was tested in a laboratory. The data collected are shown in the table.

Coats of paint (x)	1	2	3	4	5
Protection (years) (y)	1.4	2.9	4.1	5.8	7.2

a Calculate S_{xx} and S_{xy}.

b Find the equation of the regression line of y on x in the form $y = a + bx$.

7.4 Coding is sometimes used to simplify calculations.

- The numbers used in the coding to subtract and divide are selected to make the resulting numbers small.

- The coded regression line may not be the same as the actual regression line. The linear change may affect the answer.

- To carry out a linear change of variable, substitute the linear formula into the answer and simplify.

Example 5

Eight samples of carbon steel were produced with different percentages, c, of carbon in them. Each sample was heated in a furnace until it melted and the temperature, m in °C, at which it melted was recorded.

The results were coded such that $x = 10c$ and $y = \dfrac{m - 700}{5}$.

The coded results are shown in the table.

Carbon x	1	2	3	4	5	6	7	8
Melting point y	35	28	24	16	15	12	8	6

a Calculate S_{xy} and S_{xx}.

(You may use $\sum x^2 = 204$ and $\sum xy = 478$.)

b Find the regression line of y on x.

c Find the equation of the regression line of m on c.

a $S_{xy} = \sum xy - \dfrac{\sum x \sum y}{n}$

$$= 478 - \frac{36 \times 144}{8}$$

$$= -170$$

$S_{xx} = \sum x^2 - \dfrac{(\sum x)^2}{n}$

$$= 204 - \frac{36^2}{8}$$

$$= 42$$

$\sum x = 1 + 2 + 3 + 4 + 5 + 6 + 7 + 8 = 36$
$\sum y = 35 + 28 + 24 + 16 + 15 + 12 + 8 + 6 = 144$

b $b = \dfrac{S_{xy}}{S_{xx}} = \dfrac{-170}{42} = -4.048$

$a = \bar{y} - b\bar{x}$

$$= \frac{144}{8} + 4.048 \times \frac{36}{8}$$

$$= 36.216$$

$\bar{y} = \dfrac{\sum y}{n}$ and $\bar{x} = \dfrac{\sum x}{n}$

$y = 36.216 - 4.048x$

c   If $y = 36.216 - 4.048x$

$\dfrac{m - 700}{5} = 36.216 - 4.048 \times 10c$

$m - 700 = 181.08 - 202.4c$

$m = 881.08 - 202.4c$

> Substituting $\dfrac{m - 700}{5}$ for y and $10c$ for x.

> Multiply both sides by 5 and collect terms to simplify the answer.

Exercise 7B

1 Given that the coding $p = x + 2$ and $q = y - 3$ has been used to get the regression equation $p + q = 5$ find the equation of the regression line of y on x in the form $y = a + bx$.

2 Given the coding $x = p - 10$ and $y = s - 100$ and the regression equation $x = y + 2$ work out the equation of the regression line of s on p.

3 Given that the coding $g = \dfrac{x}{3}$ and $h = \dfrac{y}{4} - 2$ has been used to get the regression equation $h = 6 - 4g$ find the equation of the regression line of y on x.

4 The regression line of t on s is found by using the coding $x = s - 5$ and $y = t - 10$. The regression equation of y on x is $y = 14 + 3x$. Work out the regression line of t on s.

5 A regression line of c on d is worked out using the coding $x = \dfrac{c}{2}$ and $y = \dfrac{d}{10}$.

   a Given $S_{xy} = 120$, $S_{xx} = 240$, the mean of x ($\bar{x}$) is 5 and the mean of y ($\bar{y}$) is 6, calculate the regression line of y on x.

   b Find the regression line of d on c.

6 Some data on heights (h) and weights (w) were collected. The results were coded such that $x = \dfrac{h - 8}{2}$ and $y = \dfrac{w}{5}$. The coded results are shown in the table.

x	1	5	10	16	17
y	9	12	16	21	23

   a Calculate S_{xy} and S_{xx} and use them to find the equation of the regression line of y on x.

   b Find the equation of the regression line of w on h.

7.5 Applying and interpreting the regression equation.

- A regression line can be used to estimate the value of the dependent variable for any value of the independent variable.

- **Interpolation** is when you estimate the value of a dependent variable within the range of the data.

- **Extrapolation** is when you estimate a value outside the range of the data. Values estimated by extrapolation can be unreliable.

- You should not, in general, extrapolate and you must view any extrapolated values with caution.

Example 6

The results from an experiment in which different masses were placed on a spring and the resulting length of the spring measured, are shown below.

Mass, x (kg)	20	40	60	80	100
Length, y (cm)	48	55.1	56.3	61.2	68

The regression line for these data was found to be $y = 43.89 + 0.2305x$.

Find,

a i the value for y when the mass $x = 35$ kg,

  ii the value of y when the mass $x = 120$ kg.

b Explain what the two constants 43.89 and 0.2305 mean in this context.

The constants may have some meaning within the context of the investigation you are making, and you should be prepared to interpret them.

a i When load $x = 35$ kg

Length $y = 43.89 + 0.2305 \times 35$

$= 51.96$ cm

The value 35 kg for x lies within the range of x values. We call this **interpolation**. This is a reliable estimate.

ii When load $x = 120$ kg

$y = 43.89 + 0.2305 \times 120$

$= 71.55$ cm

The value 120 kg lies outside the range of x values. We call this extrapolation. This may not be a reliable estimate.

b 43.89 cm is the length of the spring when $x = 0$ i.e. when no load is placed on it.

0.2305 is the amount by which the spring's length increases for every extra 1 kg of load.

Example 7

A scientist working in agricultural research believes that there is a linear relationship between the amount of a certain food supplement given to hens and the hardness of the shells of the eggs they lay. As an experiment, controlled quantities of the supplement were added to the hens' normal diet for a period of two weeks and the hardness of the shells of the eggs laid at the end of this period was then measured on a scale from one to 10, with the following results:

Food supplement, f (g/day)	2	4	6	8	10	12	14
Hardness of shells, h	3.2	5.2	5.5	6.4	7.2	8.5	9.8

a Find the equation of the regression line of f on h.

b Interpret what the values of a and b tell you.

c Explain why you should not try to calculate the shell hardness for a food supplement of 20 g per day.

(You may use $\sum f = 56$, $\sum h = 45.8$, $\bar{f} = 8$, $\bar{h} = 6.543$, $\sum f^2 = 560$, $\sum fh = 422.6$)

a
$$S_{fh} = 422.6 - \frac{56 \times 45.8}{7} = 56.2$$
$$S_{ff} = 560 - \frac{56^2}{7} = 112$$
$$b = \frac{S_{fh}}{S_{ff}} = \frac{56.2}{112}$$
$$= 0.5018 \text{ hardness units per g per day}$$
$$a = \bar{h} - b\bar{f}$$
$$= 6.543 - 0.5018 \times 8$$
$$= 2.5286 \text{ hardness units}$$
$$h = 2.53 + 0.502f$$

When dealing with a real problem do not forget to put the units of measurement for the two constants.

b a tells you the shell strength when no supplement is given (i.e. when $f = 0$). Zero is only just outside the range of f so it is reasonable to use this value. b tells you the rate at which the hardness increases with increased food supplement, in this case for every extra one gram of food supplement per day the hardness increases by 0.5018 hardness units.

c The value of 20 g for f is well outside the range of f and a value for h based on this could be very unreliable.

Example 8

A repair workshop finds it is having a problem with a pressure gauge it uses.
It decides to have it checked by a specialist firm.
The following data was obtained.

Reading of Gauge x (bar)	1.0	1.4	1.8	2.2	2.6	3.0	3.4	3.8
Correct readings y (bar)	0.96	1.33	1.75	2.14	2.58	2.97	3.38	3.75

(You may use $\sum x = 19.2$ $\sum x^2 = 52.8$ $\sum y = 18.86$ $\sum y^2 = 51.30$ $\sum xy = 52.04$)

a Show that $S_{xy} = 6.776$ and find S_{xx}.

It is thought that a linear relationship of the form $y = a + bx$ could be used to describe these data.

b Use linear regression to find the value of a and b giving your answers to one decimal place.

c Draw a scatter diagram to represent these data and draw the regression line on your diagram.

d The gauge shows a reading of 2 bar. Using the regression equation work out what the correct reading would be.

a $S_{xy} = 52.04 - \dfrac{19.2 \times 18.86}{8} = 6.776$

    $S_{xx} = 52.8 - \dfrac{19.2 \times 19.2}{8} = 6.72$

b $b = \dfrac{6.776}{6.762} = 1.0083\ldots$

    $a = \dfrac{18.86}{8} - 1.0083\ldots \times \dfrac{19.2}{8}$

      $= -0.0625$

    Regression line is: $y = -0.0625 + 1.01x$

    or $y = 1.01x - 0.0625$

c

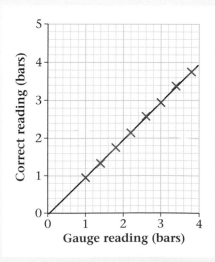

To draw the regression line either plot the point (0, a) and use the gradient as in Example 1 or find two points on the line.

In this case using $x = 1$ gives $y = 0.95$ and using $x = 3$ gives $y = 2.96$.

d $y = (1.0083 \times 2) - 0.0625$

    $y = 1.95$ bar

Exercise 7C

1 Given the regression line $y = 24 - 3x$ find the value of y when x is 6.

2 The regression line for the weight (w) in grams on the volume (v) in cm^3 for a sample of small marbles is $w = 300 + 12v$.
Calculate the weight when the volume is $7\,\text{cm}^3$.

3 **a** State what is meant by extrapolation.

   b State what is meant by interpolation.

4 12 children between the ages (x) of five and 11 years were asked how much pocket money (y) they were given each week. The equation for the regression line of y on x was found to be $y = 2x - 8$.

   a Use the equation to estimate the amount of money a seven year old would get. State, with a reason, whether or not this is a reliable estimate.

   b One of the children suggested that this equation must be wrong since it showed that a three year old would get a negative amount of pocket money. Explain why this has happened.

5 The pulse rates (y) of 10 people were measured after doing different amounts of exercise (x) for between two and 10 minutes. The regression equation $y = 0.75x + 72$ refers to these data. The equation seems to suggest that someone doing 30 minutes of exercise would have a pulse rate of 94.5. State whether or not this is sensible. Give a reason for your answer.

6 Over a period of time the sales, (y) in thousands, of 10 similar text books and the amount, (x) in £ thousands, spent on advertising each book were recorded. The greatest amount spent on advertising was £4.4 thousand, and the least amount was £0.75 thousand.

An equation of the regression line for y on x was worked out for the data.

The equation was $y = 0.93 + 1.1x$.

   a Use the equation to estimate the sales of a text book if the amount spent on advertising is to be set at £2.65 thousand. State, with a reason, whether or not this is a reliable estimate.

   b Use the equation to estimate the sales of the book if the amount to be spent on advertising is £8000. State, with a reason, whether or not this is a reliable estimate.

   c Explain what the value 1.1 tells you about the relationship between the sales of books and the amount spent on advertising.

   d Interpret the meaning of the figure 0.93.

7 Research was done to see if there is a relationship between finger dexterity and the ability to do work on a production line. The data is shown in the table.

Dexterity Score (x)	2.5	3	3.5	4	5	5	5.5	6.5	7	8
Productivity (y)	80	130	100	220	190	210	270	290	350	400

The equation of the regression line for these data is $y = -59 + 57x$

a Use the equation to estimate the productivity of someone with a dexterity of 6.

b State the contextual meaning of the figure 57.

c State, giving in each case a reason, whether or not it would be reasonable to use this equation to work out the productivity of someone with dexterity of:

  i 2   ii 14.

8 A regression line of $y = 5 + 3x$ is found using 10 data sets. Another piece of data is recorded which when put on a scatter diagram with the original data proves to be well above the regression line for all the other data. Write down whether or not the regression equation would change if this piece of data were included in the calculation.

Mixed exercise 7D

1 A metal rod was found to increase in length as it was heated. The temperature (t) and the increase in length (l mm) were measured at intervals between 30°C and 400°C degrees. The regression line of l on t was found to be $l = 0.009t - 0.25$.

a Find the increase in length for a temperature of 300°C.

b Find the increase in length for a temperature of 530°C.

c Write down, with reasons, why the answer to **a** might be reliable and the answer **b** unreliable.

2 Two variables s and t are thought to be connected by a law of the form $t = a + bs$, where a and b are constants.

a Use the summary data:

$\sum s = 553$ $\sum t = 549$ $\sum st = 31\,185$ $n = 12$ $\bar{s} = 46.0833$
$\bar{t} = 45.75$ $S_{ss} = 6193$

to work out the regression line of t on s.

b Find the value of t when s is 50

3 A biologist recorded the breadth (x cm) and the length (y cm) of 12 beech leaves. The data collected can be summarised as follows.

$\sum x^2 = 97.73$ $\sum x = 33.1$ $\sum y = 66.8$ $\sum xy = 195.94$

a Calculate S_{xx} and S_{xy}.

b Find the equation of the regression line of y on x in the form $y = a + bx$.

c Predict the length of a beech leaf that has a breadth of 3.0 cm.

4 Energy consumption is claimed to be a good predictor of Gross National Product. An economist recorded the energy consumption (x) and the Gross National Product (y) for eight countries. The data are shown in the table.

Energy Consumption x	3.4	7.7	12.0	75	58	67	113	131
Gross National Product y	55	240	390	1100	1390	1330	1400	1900

a Calculate S_{xy} and S_{xx}.

b Find the equation of the regression line of y on x in the form $y = a + bx$.

c Estimate the Gross National Product of a country that has an energy consumption of 100.

d Estimate the energy consumption of a country that has a Gross National Product of 3500.

e Comment on the reliability of your answer to **d**.

5 In an environmental survey on the survival of mammals the tail length t (cm) and body length m (cm) of a random sample of six small mammals of the same species were measured. These data are coded such that $x = \frac{m}{2}$ and $y = t - 2$.

The data from the coded records are summarised below.

$$\sum y = 13.5 \qquad \sum x = 25.5 \qquad \sum xy = 84.25 \qquad S_{xx} = 59.88$$

a Find the equation of the regression line of y on x in the form $y = ax + b$.

b Hence find the equation of the regression line of t on m.

c Predict the tail length of a mammal that has a body length of 10 cm.

6 A health clinic counted the number of breaths per minute (r) and the number of pulse beats (p) per minute for 10 people doing various activities. The data are shown in the table.

The data are coded such that $x = \frac{r - 10}{2}$ and $y = \frac{p - 50}{2}$.

x	3	5	5	7	8	9	9	10	12	13
y	4	9	10	11	17	15	17	19	22	27

(You may use $\sum x = 81 \quad \sum x^2 = 747 \quad \sum y = 151 \quad \sum y^2 = 2695 \quad \sum xy = 1413$.)

a Calculate S_{xy} and S_{xx}.

b Find the equation of the regression line of y on x in the form $y = a + bx$.

c Find the equation of the regression line for p on r.

d Estimate the number of pulse beats per minute for someone who is taking 22 breaths per minute.

e Comment on the reliability of your answer to **d**.

7 A farm food supplier monitors the number of hens kept (x) against the weekly consumption of hen food (y kg) for a sample of 10 small holders. He records the data and works out the regression line for y on x to be $y = 0.16 + 0.79x$.

a Write down a practical interpretation of the figure 0.79.

b Estimate the amount of food that is likely to be needed by a small holder who has 30 hens.

c If food costs £12 for a 10 kg bag estimate the weekly cost of feeding 50 hens.

8 Water voles are becoming very rare; they are often confused with water rats. A naturalist society decided to record details of the water voles in their area. The members measured the weight (y) to the nearest 10 grams, and the body length (x) to the nearest millimetre, of eight active healthy water voles. The data they collected are in the table.

Body Length (x) mm	140	150	170	180	180	200	220	220
Weight (y) grams	150	180	190	220	240	290	300	310

a Draw a scatter diagram of these data.

b Give a reason to support the calculation of a regression line for these data.

c Use the coding $l = \dfrac{x}{10}$ and $w = \dfrac{y}{10}$ to work out the regression line of w on l.

d Find the equation of the regression line for y on x.

e Draw the regression line on the scatter diagram.

f Use your regression line to calculate an estimate for the weight of a water vole that has a body length of 210 mm. Write down, with a reason, whether or not this is a reliable estimate.

The members of the society remove any water voles that seem unhealthy from the river and take them into care until they are fit to be returned.

They find three water voles on one stretch of river which have the following measurements.

A: Weight 235 gm and body length 180 mm
B: Weight 180 gm and body length 200 mm
C: Weight 195 gm and body length 220 mm

g Write down, with a reason, which of these water voles were removed from the river.

9 A mail order company pays for postage of its goods partly by destination and partly by total weight sent out on a particular day. The number of items sent out and the total weights were recorded over a seven day period. The data are shown in the table.

Number of items (n)	10	13	22	15	24	16	19
Weight in kg (w)	2800	3600	6000	3600	5200	4400	5200

a Use the coding $x = n - 10$ and $y = \dfrac{w}{400}$ to work out S_{xy} and S_{xx}.

b Work out the equation of the regression line for y on x.

c Work out the equation of the regression line for w on n.

d Use your regression equation to estimate the weight of 20 items.

e State why it would be unwise to use the regression equation to estimate the weight of 100 items.

Summary of key points

1   If $y = a + bx$ then a (sometimes called the intercept) is where the line cuts the y-axis (the line $x = 0$) and b is the amount by which y increases for an increase of 1 in x, (b is called the gradient of the line).

2   An independent (or explanatory) variable is one that is set independently of the other variable.

   A dependent (or response) variable is one whose values are decided by the values of the independent variable.

3   The equation of the regression line of y on x is:

   $y = a + bx$

   where $b = \dfrac{S_{xy}}{S_{xx}}$ and $a = \bar{y} - b\bar{x}$.

4   Coding is sometimes used to simplify calculations.

   To turn a coded regression line into an actual regression line you substitute the codes into the answer.

5   Interpolation is when you estimate the value of a dependent variable within the range of the data.

6   Extrapolation is when you estimate a value outside the range of the data.

   Values estimated by extrapolation can be unreliable.

8

After completing this chapter you should be able to

- understand what a **discrete random variable** is
- understand how discrete random variables arise
- be able to find the **cumulative distribution function** of a discrete random variable
- find the **mean** and **variance** of a discrete random variable
- be able to use the **discrete uniform distribution**.

Discrete random variables

In a game a contestant is given the chance to choose one of five plain brown envelopes but must pay £10 for this chance. One envelope contains a blank piece of paper, one a £5 note, one a £10 note, one a £20 note and the last one a £50 note. What is the mean of his expected winnings, and is this a fair game? You will be able to answer this question when you have studied this chapter.

8.1 A discrete random variable can be described as a value obtained by taking a measurement from an experiment in the real world. A random variable must take a numerical value.

- A **variable** is represented by a symbol (X, Y, A, B etc.), and it can take on any of a specified set of values.

- When the value of a variable is the outcome of an experiment, the variable is called a **random variable**.

- Another name for the list of all possible outcomes of an experiment is the **sample space**.

Sometimes you may want to differentiate between a random variable X and one of its values.

- Capital letters like X are used for the random variable and a small letter such as x for a particular value of the random variable X.

- The probability that X is equal to a particular value x, is written as $P(X = x)$ or sometimes as $p(x)$. The two notations are freely interchangeable.

For a random variable X:

- x is a particular value of X.

- $P(X = x)$ refers to the probability that X is equal to a particular value x.

Random variables may be discrete or continuous.

- A continuous random variable is one where the outcome can be any value on a continuous scale. (You will meet a continuous variable in Chapter 9.)

- A discrete random variable takes only values on a discrete scale.

This chapter is concerned only with discrete random variables.

Example 1

A coin is tossed six times and the number of heads, X, is noted.
Write down all possible values of X.

Possible values of X are
$x = 0, 1, 2, 3, 4, 5$ or 6.

Tossing the coin is an experiment.
The possible values of X are the possible outcomes of the experiment.

Example 2

Write down whether or not each of the following is a discrete random variable.

a The average height of a group of boys.

b The number of times a coin is tossed before a head appears.

c The number of months in a year.

a	Is not a discrete random variable.
b	Is a discrete random variable.
c	Is not a discrete random variable.

Height is measured on a continuous scale.

It is a number that is the result of an experiment.

It does not vary and is not the result of an experiment.

8.2 **A discrete random variable can be specified by giving the set of all possible outcomes of the experiment or the possible values of the random variable and the probability with which each occurs.**

■ To specify a discrete random variable completely, you need to know its set of possible values and the probability with which it takes each one.

■ You can draw up a table to show the probability of each outcome of an experiment. This is called a **probability distribution**.

■ You can also specify a discrete random variable as a function. For example the random variable in Example 3 can be written as:

$$P(X = x) = \frac{1}{6} \text{ for } x = 1, 2, 3, 4, 5, 6.$$

This is known as a **probability function**. (See Example 4.)

Example 3

A fair die is rolled. Write down the probability of getting any particular number.

x	1	2	3	4	5	6
$P(X = x)$	$\frac{1}{6}$	$\frac{1}{6}$	$\frac{1}{6}$	$\frac{1}{6}$	$\frac{1}{6}$	$\frac{1}{6}$

These are the possible values.

These are the probabilities of each value.

Example 4

Three fair coins are tossed. The number of heads, X, is counted.
For this experiment:

a write down the sample space,   **b** write down the probability distribution,

c write down the probability function.

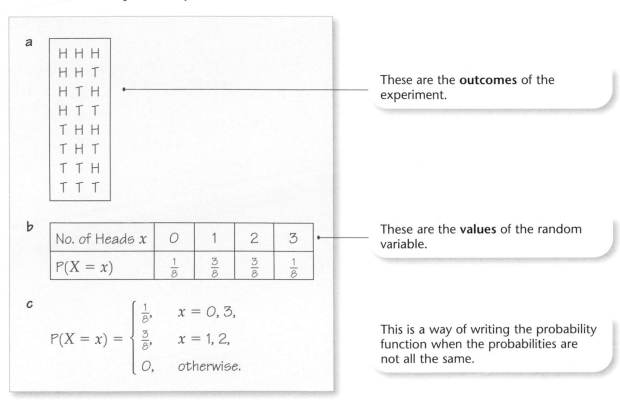

a

H H H
H H T
H T H
H T T
T H H
T H T
T T H
T T T

These are the **outcomes** of the experiment.

b

No. of Heads x	0	1	2	3
$P(X = x)$	$\frac{1}{8}$	$\frac{3}{8}$	$\frac{3}{8}$	$\frac{1}{8}$

These are the **values** of the random variable.

c

$$P(X = x) = \begin{cases} \frac{1}{8}, & x = 0, 3, \\ \frac{3}{8}, & x = 1, 2, \\ 0, & \text{otherwise.} \end{cases}$$

This is a way of writing the probability function when the probabilities are not all the same.

8.3 **For a discrete random variable the sum of all the probabilities must add up to one.**

■ In both Example 3 ($\frac{1}{6} + \frac{1}{6} + \frac{1}{6} + \frac{1}{6} + \frac{1}{6} + \frac{1}{6} = 1$) and Example 4 ($\frac{1}{8} + \frac{3}{8} + \frac{3}{8} + \frac{1}{8} = 1$) the sum of all the probabilities was 1.

■ Or, in symbols, $\sum p(x) = \sum P(X = x) = 1$ for all values of x.

This is an important property of discrete random variables, and you may have to use this property in an examination question.

Example 5

A tetrahedral die has the numbers 1, 2, 3 and 4 on its faces. The die is biased in such a way that the probability of the die landing on any number x is $\frac{k}{x}$, where k is a constant.

Find the probability distribution for X, the number the die lands on after a single roll.

The probability distribution will be:

x	1	2	3	4
$P(X = x)$	$\frac{k}{1}$	$\frac{k}{2}$	$\frac{k}{3}$	$\frac{k}{4}$

Since this is a probability distribution

$$\sum P(X = x) = 1$$

$$\frac{k}{1} + \frac{k}{2} + \frac{k}{3} + \frac{k}{4} = 1$$

Therefore $k(1 + \frac{1}{2} + \frac{1}{3} + \frac{1}{4}) = 1$.

$$k\left(\frac{12 + 6 + 4 + 3}{12}\right) = 1$$

$$k = \frac{12}{25}$$

The probability distribution is:

x	1	2	3	4
$P(X = x)$	$\frac{12}{25}$	$\frac{6}{25}$	$\frac{4}{25}$	$\frac{3}{25}$

Here $\sum P(X = x) = 1$ is being used to find the value of the constant k.

Exercise 8A

1 Write down whether or not each of the following is a discrete random variable. Give a reason for your answer.

   a The average lifetime of a battery.

   b The number of days in a week.

   c The number of moves it takes to complete a game of chess.

2 A fair die is thrown four times and the number of times it falls with a 6 on the top, Y, is noted. Write down all the possible values of y.

3 A bag contains two discs with the number 2 on them and two discs with the number 3 on them. A disc is drawn at random from the bag and the number noted. The disc is returned to the bag. A second disc is then drawn from the bag and the number noted.

   a Write down the sample space.

   The discrete random variable X is the sum of the two numbers.

   b Write down the probability distribution for X.

   c Write down the probability function for X.

4 A discrete random variable X has the following probability distribution:

x	1	2	3	4
$P(X = x)$	$\frac{1}{3}$	$\frac{1}{3}$	k	$\frac{1}{4}$

Find the value of k.

5 The random variable X has a probability function

$$P(X = x) = kx \qquad x = 1, 2, 3, 4.$$

Show that $k = \frac{1}{10}$.

6 The random variable X has the probability function

$$P(X = x) = \frac{x - 1}{10} \qquad x = 1, 2, 3, 4, 5.$$

Construct a table giving the probability distribution of X.

7 The random variable X has a probability function

$$P(X = x) = \begin{cases} kx & x = 1, 3 \\ k(x - 1) & x = 2, 4 \end{cases}$$

where k is a constant.

a Find the value of k.

b Construct a table giving the probability distribution of X.

8 The discrete random variable X has a probability function

$$P(X = x) = \begin{cases} 0.1 & x = -2, -1 \\ \beta & x = 0, 1 \\ 0.2 & x = 2 \end{cases}$$

a Find the value of β.

b Construct a table giving the probability distribution of X.

9 A discrete random variable has the probability distribution shown in the table below.

x	0	1	2
$P(X = x)$	$\frac{1}{4} - a$	a	$\frac{1}{2} + a$

Find the value of a.

8.4 Finding the probability that X is less than a particular value, greater than a particular value or lies between two values.

Example 6

A discrete random variable X has the following probability distribution:

x	1	2	3	4	5	6
$P(X = x)$	0.1	0.2	0.3	0.25	0.1	0.05

Find

a $P(1 < X < 5)$,

b $P(2 \leqslant X \leqslant 4)$,

c $P(3 < X \leqslant 6)$,

d $P(X < 3)$.

a $P(1 < X < 5) = P(X = 2, 3, \text{ or } 4)$

$\qquad\qquad\quad = 0.2 + 0.3 + 0.25$

$\qquad\qquad\quad = 0.75$

If $x < 5$ then $x \leqslant 4$.

b $P(2 \leqslant X \leqslant 4) = P(X = 2, 3 \text{ or } 4)$

$\qquad\qquad\quad = 0.75$

c $P(3 < X \leqslant 6) = P(X = 4, 5 \text{ or } 6)$

$\qquad\qquad\quad = 0.25 + 0.1 + 0.05$

$\qquad\qquad\quad = 0.4$

If $3 < x$ then $4 \leqslant x$.

d $P(X \leqslant 3) \quad = P(X = 1, 2 \text{ or } 3)$

$\qquad\qquad\quad = 0.1 + 0.2 + 0.3$

$\qquad\qquad\quad = 0.6$

8.5 Finding the cumulative distribution function for a discrete random variable.

■ If a particular value of X is x, the probability that X is less than or equal to x, is written as $F(x)$. $F(x)$ is found by adding together all the probabilities for those outcomes that are equal to or less than x. This is written as:

$\qquad F(x) = P(X \leqslant x)$

■ Like a probability distribution, a **cumulative distribution function** can be written as a table.

Example 7

Two fair coins are tossed. X is the number of heads showing on the two coins.
Draw up a table to show the cumulative distribution function for X.

The sample space is

HH HT TH TT

No. of Heads x	0	1	2
$P(X = x)$	0.25	0.5	0.25
$F(x)$	0.25	0.75	1

$F(0) = p(0) = 0.25$
$F(1) = p(0) + p(1) = 0.25 + 0.5 = 0.75$
$F(2) = p(0) + p(1) + p(2)$ or use $F(1) + p(2)$
$\quad\quad = 0.25 + 0.5 + 0.25 = 1$

Example 8

The discrete random variable X has a cumulative distribution function $F(x)$ defined by

$$F(x) = \frac{(x + k)}{8}; x = 1, 2 \text{ and } 3.$$

a Find the value of k.

b Draw the distribution table for the cumulative distribution function.

c Write down $F(2.6)$

d Find the probability distribution of X.

a $F(3) = 1$ so $\dfrac{3 + k}{8} = 1$

$3 + k = 8$

$k = 5$

All values of x are equal to or less than 3, so $F(3) = 1$ since all probabilities must add to 1.

b $F(2) = \dfrac{2 + 5}{8} = \dfrac{7}{8}$

$F(1) = \dfrac{1 + 5}{8} = \dfrac{3}{4}$

x	1	2	3
$F(x)$	$\dfrac{3}{4}$	$\dfrac{7}{8}$	1

c $F(2.6) = F(2) = \dfrac{7}{8}$

$F(2.6)$ means $P(X \leqslant 2.6)$ but X doesn't take any values between 2 and 3 so $X \leqslant 2.6$ is the same as $X \leqslant 2$ and thus $F(2.6) = F(2)$. Remember X is a discrete random variable so you do not interpolate.

d $p(1) = F(1) = \dfrac{3}{4}$

$p(2) = F(2) - F(1) = \dfrac{7}{8} - \dfrac{3}{4} = \dfrac{1}{8}$

$p(3) = F(3) - F(2) = 1 - \dfrac{7}{8} = \dfrac{1}{8}$

The distribution is

x	1	2	3
$P(X = x)$	$\dfrac{3}{4}$	$\dfrac{1}{8}$	$\dfrac{1}{8}$

Exercise 8B

1 A discrete random variable X has probability distribution

x	0	1	2	3	4	5
$P(X = x)$	0.1	0.1	0.3	0.3	0.1	0.1

  a Find the probability that $X < 3$.
  b Find the probability that $X > 3$.
  c Find the probability that $1 < X < 4$.

2 A discrete random variable X has probability distribution

x	0	1	2	3
$P(X = x)$	$\dfrac{1}{8}$	$\dfrac{1}{4}$	$\dfrac{1}{2}$	$\dfrac{1}{8}$

  Find
  a $P(1 < X \leqslant 3)$, **b** $P(X < 2)$.

3 A discrete random variable X has a probability distribution

x	1	2	3	4	5	6
$P(X = x)$	0.1	0.1	0.15	0.25	0.3	0.1

  a Draw up a table to show the cumulative distribution function $F(x)$.
  b Write down $F(5)$. **c** Write down $F(2.2)$.

4 A discrete random variable has a cumulative distribution function $F(x)$ given in the table.

x	0	1	2	3	4	5	6
$F(x)$	0	0.1	0.2	0.45	0.5	0.9	1

  a Draw up a table to show the probability distribution X.
  b Write down $P(X < 5)$. **c** Find $P(2 \leqslant X < 5)$.

5 The random variable X has a probability function

$$P(X = x) = \begin{cases} kx & x = 1, 3, 5 \\ k(x - 1) & x = 2, 4, 6 \end{cases}$$

where k is a constant.

a Find the value of k.

b Draw a table giving the probability distribution of X.

c Find $P(2 \leqslant X < 5)$.

d Find $F(4)$.

e Find $F(1.6)$.

6 The discrete random variable X has the probability function

$$P(X = x) = \begin{cases} 0.1 & x = -2, -1 \\ \alpha & x = 0, 1 \\ 0.3 & x = 2 \end{cases}$$

a Find the value of α.

b Draw a table giving the probability distribution of X.

c Write down the value of $F(0.3)$.

7 The discrete random variable X has a cumulative distribution function $F(x)$ defined by

$$F(x) = \begin{cases} 0 & x = 0 \\ \dfrac{1 + x}{6} & x = 1, 2, 3, 4, 5 \\ 1 & x > 5 \end{cases}$$

a Find $P(X \leqslant 4)$.

b Show that $P(X = 4)$ is $\frac{1}{6}$.

c Find the probability distribution for X.

8 The discrete random variable X has a cumulative distribution function $F(x)$ defined by

$$F(x) = \begin{cases} 0 & x = 0 \\ \dfrac{(x + k)^2}{16} & x = 1, 2 \text{ and } 3 \\ 1 & x > 3 \end{cases}$$

a Find the value of k.

b Find the probability distribution for X.

8.6 **Finding a mean or expected value of a discrete random variable.**

■ The expected value of X is defined as:

expected value of $X = E(X) = \sum x P(X = x) = \sum x p(x)$

Remember $p(x)$ is the same as $P(X = x)$.

■ If you have a statistical experiment:
a practical approach results in a frequency distribution and a mean value, a theoretical approach results in a probability distribution and an expected value.

Example 9

The number of television sets per household in a survey of a 100 houses in a town gave the following frequency distribution:

Number of sets	0	1	2	3
Frequency	10	75	10	5

a Find the mean of this set of data.

b Draw up the probability distribution table for the variable X where X is the number of TV sets in a house picked at random in the town.

c Find the expected value for X.

a mean $= \dfrac{\sum fx}{\sum f} = \dfrac{(0 \times 10) + (1 \times 75) + (2 \times 10) + (3 \times 5)}{100}$

$= (0 \times 0.1) + (1 \times 0.75) + (2 \times 0.1) + (3 \times 0.05)$

$= 0.75 + 0.2 + 0.15 = 1.1$

b

x	0	1	2	3
$p(x)$	0.1	0.75	0.1	0.05

Probability $= \dfrac{f}{\sum f}$.

c $E(X) = \sum xp(x)$

$= (0 \times 0.1) + (1 \times 0.75) + (2 \times 0.1) + (3 \times 0.05)$

$= 0.75 + 0.2 + 0.15 = 1.1$

In Example 9 mean $= \dfrac{\sum fx}{\sum f} = 1.1$ and $E(X) = \sum xp(x) = 1.1$

The calculations for the mean and the $E(X)$ are similar.
For this reason $E(X)$ is sometimes called the mean of X.

Example 10

The random variable X has a probability distribution

a Given that $E(X) = 3$, write down two equations involving p and q.

b Find the value of p and the value of q.

x	1	2	3	4	5
$p(x)$	0.1	p	0.3	q	0.2

a $p + q + 0.1 + 0.3 + 0.2 = 1$ — $\sum P(X = x) = 1$

$p + q = 1 - 0.6$

$p + q = 0.4 \qquad (1)$

$(1 \times 0.1) + 2p + (3 \times 0.3) + 4q + (5 \times 0.2) = 3$ — $E(X) = \sum xP(X = x)$

$2p + 4q = 3 - (0.1 + 0.9 + 1)$

$2p + 4q = 1 \qquad (2)$

b $2p + 4q = 1$ ———— From (2).

$2p + 2q = 0.8$ ———— Multiply (1) by 2.

so $2q = 0.2$ ———— Subtract bottom line from top line.

$q = 0.1$

$p = 0.4 - q$

$= 0.4 - 0.1$ ———— Substitute 0.1 for q.

$= 0.3$

8.7 Finding an expected value for X^2.

■ $E(X^2) = \sum x^2 P(X = x)$

■ This is an example of a more general rule: $E(X^n) = \sum x^n P(X = x)$

Example 11

A discrete random variable X has a probability distribution

x	1	2	3	4
$P(X = x)$	$\frac{12}{25}$	$\frac{6}{25}$	$\frac{4}{25}$	$\frac{3}{25}$

a Write down the probability distribution for X^2.

b Find $E(X^2)$.

a The distribution for X^2 is

x	1	2	3	4
x^2	1	4	9	16
$P(X = x^2)$	$\frac{12}{25}$	$\frac{6}{25}$	$\frac{4}{25}$	$\frac{3}{25}$

Note $P(X = x^2) = P(X = x)$.

b $E(X^2) = \sum x^2 P(X = x^2)$

$= 1 \times \frac{12}{25} + 4 \times \frac{6}{25} + 9 \times \frac{4}{25} + 16 \times \frac{3}{25}$

$= \frac{120}{25}$

$= 4.8$

Note $E(X)$ is actually $= 1.92$. So $E(X)^2 \neq [E(X)]^2$.

1 Find $E(X)$ and $E(X^2)$ for the following distributions of x.

a

x	2	4	6	8
$P(X = x)$	0.3	0.3	0.2	0.2

b

x	1	2	3	4
$P(X = x)$	0.1	0.4	0.1	0.4

2 A biased die has the probability distribution

x	1	2	3	4	5	6
$P(X = x)$	0.1	0.1	0.1	0.2	0.4	0.1

Find $E(X)$ and $E(X^2)$.

3 The random variable X has a probability function

$$P(X = x) = \begin{cases} \dfrac{1}{x} & x = 2, 3, 6 \\ 0 & \text{all other values} \end{cases}$$

a Construct a table giving the probability distribution of X.

b Work out $E(X)$ and $E(X^2)$.

c State with a reason whether or not $(E(X))^2 = E(X^2)$.

4 Two coins are tossed 50 times. The number of heads is noted each time.

a Construct a probability distribution table for the number of heads when the two coins are tossed once, assuming that the two coins are unbiased.

b Work out how many times you would expect to get 0, 1 or 2 heads.

The following table shows the actual results.

Number of heads (h)	0	1	2
Frequency (f)	7	22	21

c State whether or not the actual frequencies obtained support the statement that the coins are unbiased. Give a reason for your answer.

5 The random variable X has the following probability distribution.

x	1	2	3	4	5
$P(X = x)$	0.1	a	b	0.2	0.1

Given $E(X) = 2.9$ find the value of a and the value of b.

6 A fair spinner with equal sections numbered 1 to 5 is thrown 500 times. Work out how many times it can be expected to land on 3.

8.8 Finding the variance of a random variable.

■ The variance of X is usually written as Var(X) and is found by using:

$$\text{Var}(X) = E(X^2) - (E(X))^2$$

Example 12

Two fair cubical dice are thrown: one is red and one is blue. The random variable M represents the score on the red die minus the score on the blue die.

a Find the probability distribution of M.

b Write down $E(M)$.

c Find Var(M).

a You can represent the value of M in the following table:

Blue Red	1	2	3	4	5	6
1	0	−1	−2	−3	−4	−5
2	1	0	−1	−2	−3	−4
3	2	1	0	−1	2	−3
4	3	2	1	0	−1	−2
5	4	3	2	1	0	−1
6	5	4	3	2	1	0

Probabilities can be worked out from the number of times each number appears in the table.

m	−5	−4	−3	−2	−1	0	1	2	3	4	5
$P(M = m)$	$\frac{1}{36}$	$\frac{2}{36}$	$\frac{3}{36}$	$\frac{4}{36}$	$\frac{5}{36}$	$\frac{6}{36}$	$\frac{5}{36}$	$\frac{4}{36}$	$\frac{3}{36}$	$\frac{2}{36}$	$\frac{1}{36}$

b $E(M) = (-5 \times \frac{1}{36}) + (-4 \times \frac{2}{36}) + (-3 \times \frac{3}{36}) + (-2 \times \frac{4}{36})$

$\qquad + (-1 \times \frac{5}{36}) + (1 \times \frac{5}{36}) + (2 \times \frac{4}{36}) + (3 \times \frac{3}{36})$

$\qquad + (4 \times \frac{2}{36}) + (5 \times \frac{1}{36})$

$\quad = 0$ ●————————————————

(alternatively by symmetry $E(M) = 0$).

c $\text{Var}(M) = \sum m^2 P(M = m) - 0^2 = E(x)^2 - 0^2$

$$= 25 \times \frac{1}{36} + 16 \times \frac{2}{36} + 9 \times \frac{3}{36} + 4 \times \frac{4}{36} + 1 \times \frac{5}{36}$$

$$+ 0 \times \frac{6}{36} + 1 \times \frac{5}{36} + \dots + 25 \times \frac{1}{36}$$

$$= \frac{(25 + 32 + 27 + 16 + 5) \times 2}{36}$$

$$= \frac{105}{18}$$

$$= \frac{35}{6}$$

Exercise 8D

1 For the following probability distribution

x	-1	0	1	2	3
$P(X = x)$	$\frac{1}{5}$	$\frac{1}{5}$	$\frac{1}{5}$	$\frac{1}{5}$	$\frac{1}{5}$

a write down $E(X)$,

b find $\text{Var}(X)$.

2 Find the expectation and variance of each of the following distributions of X.

a

x	1	2	3
$P(X = x)$	$\frac{1}{3}$	$\frac{1}{2}$	$\frac{1}{6}$

b

x	-1	0	1
$P(X = x)$	$\frac{1}{4}$	$\frac{1}{2}$	$\frac{1}{4}$

c

x	-2	-1	1	2
$P(X = x)$	$\frac{1}{3}$	$\frac{1}{3}$	$\frac{1}{6}$	$\frac{1}{6}$

3 Given that Y is the score when a single unbiased die is rolled, find $E(Y)$ and $\text{Var}(Y)$.

4 Two fair cubical dice are rolled and S is the sum of their scores.

a Find the distribution of S.

b Find $E(S)$.

c Find $\text{Var}(S)$.

5 Two fair tetrahedral (four-sided) dice are rolled and D is the difference between their scores.

   a Find the distribution of D and show that $P(D = 3) = \frac{1}{8}$.

   b Find $E(D)$.

   c Find $Var(D)$.

6 A fair coin is tossed repeatedly until a head appears or three tosses have been made. The random variable T represents the number of tosses of the coin.

   a Show that the distribution of T is

t	1	2	3
$P(T = t)$	$\frac{1}{2}$	$\frac{1}{4}$	$\frac{1}{4}$

   b Find the expectation and variance of T.

7 The random variable X has the following distribution:

x	1	2	3
$P(X = x)$	a	b	a

   where a and b are constants.

   a Write down $E(X)$.

   b Given that $Var(X) = 0.75$, find the values of a and b.

8.9 Extending the idea of the expected value of X and the variance of X to deal with the expected value and variance of a function of X.

- If X is a random variable, then the value of X is a number determined by an experiment.

- Each function of a **random variable** has a **probability distribution** that you can derive from the probability distribution of X.

- A general rule relating $E(X)$ and $E(aX + b)$ where a and b are constants is

   $E(aX + b) = aE(X) + b.$

- In the same way $Var(X)$ and $Var(aX + b)$ are related by the rule

   $Var(aX + b) = a^2\,Var(X).$

- The **variance** is a measure of spread relative to the mean so adding a constant value to all the values of X will not affect this measure of spread hence the 'b' does not change the variance. The 'a^2' is understandable if you remember that;

   $Var(X) = E(X^2) - [E(X)]^2$

   so if each value of X is multiplied by the value of a the variance will be multiplied by a^2.

Example 13

A discrete random variable X has a probability distribution

x	1	2	3	4
P($X = x$)	$\frac{12}{25}$	$\frac{6}{25}$	$\frac{4}{25}$	$\frac{3}{25}$

a Write down the probability distribution for Y where $Y = 2X + 1$.

b Find E(Y).

a The distribution for Y is

x	1	2	3	4
y	3	5	7	9
P($Y = y$)	$\frac{12}{25}$	$\frac{6}{25}$	$\frac{4}{25}$	$\frac{3}{25}$

When $x = 1$, $y = 2 \times 1 + 1 = 3$
$x = 2$, $y = 2 \times 2 + 1 = 5$
etc.

Notice how the probabilities relating to X are still being used, for example, P($X = 3$) = P($Y = 7$).

b
$$E(Y) = E(2X + 1) = \sum y P(Y = y)$$
$$= \sum (2x + 1)p(x)$$
$$= 3 \times \frac{12}{25} + 5 \times \frac{6}{25} + 7 \times \frac{4}{25} + 9 \times \frac{3}{25}$$
$$= \frac{121}{25}$$
$$= 4.84$$

Example 14

A random variable X has E(X) = 4 and Var (X) = 3.

Find

a E($3X$),   **b** E($X - 2$),

c Var($3X$),   **d** Var ($X - 2$),

e E(X^2).

a E($3X$) = 3E(X) = 3 × 4 = 12

b E($X - 2$) = E(X) − 2 = 4 − 2 = 2

c Var($3X$) = 3^2 Var(X) = 9 × 3 = 27

d Var ($X - 2$) = Var(X) = 3

e E(X^2) = Var(X) + [E(X)]^2 = 3 + 4^2 = 19

Example 15

Two fair 10p coins are tossed. The random variable X represents the total value of the coins that land heads up.

a Find $E(X)$ and $Var(X)$.

The random variables S and T are defined as follows:

$S = X - 10$ and $T = \frac{1}{2}X - 5$

b Show that $E(S) = E(T)$.

c Find $Var(S)$ and $Var(T)$.

Susan and Thomas play a game using two 10p coins. The coins are tossed and Susan records her score using the random variable S and Thomas uses the random variable T. After a large number of tosses they compare their scores.

d Comment on any likely differences or similarities.

---

a The distribution of X is:

x	0	10	20
$P(X = x)$	$\frac{1}{4}$	$\frac{1}{2}$	$\frac{1}{4}$

$E(X) = 10$ by symmetry.

$Var(X) = E(X^2) - (E(X))^2$

$Var(X) = 0^2 \times \frac{1}{4} + 10^2 \times \frac{1}{2} + 20^2 \times \frac{1}{4} - 10^2 = 50$

b $E(S) = E(X - 10) = E(X) - 10 = 10 - 10 = 0$

$E(T) = E\left(\frac{1}{2}X - 5\right) = \frac{1}{2}E(X) - 5 = \frac{1}{2} \times 10 - 5 = 0$

c $Var(S) = Var(X) = 50$

$Var(T) = \left(\frac{1}{2}\right)^2 Var(X) = \frac{50}{4} = 12.5$

d Their total scores should both be approximately zero.

Susan's scores should be more varied than Thomas's.

Example 16

A random variable Y has mean 3 and variance 8.

Find

a $E(12 - 3Y)$

b $Var(12 - 3Y)$

a	$E(12 - 3Y) = 12 - 3E(Y)$
	$= 12 - 3 \times 3$
	$= 3$
b	$Var(12 - 3Y) = (-3)^2 Var(Y)$
	$= 9Var(Y)$
	$= 9 \times 8$
	$= 72$

8.10 **A theoretical approach to finding the mean and standard deviation of the original data given the mean and standard deviation of the coded data.**

Example 17

Data are coded using $Y = \dfrac{X - 150}{50}$. The mean of the coded data is 5.1.

The standard deviation of the coded data is 2.5.

Find

a the mean of the original data,

b the standard deviation of the original data.

a	$Y = \dfrac{X - 150}{50}$
	$X = 50Y + 150$
	$E(X) = E(50Y + 150)$
	$= 50E(Y) + 150$
	$= 255 + 150$
	$= 405$

b	$\text{Var}(X) = \text{Var}(50Y + 150)$
	$= 50^2\text{Var}(Y)$
	$= 50^2 \times 2.5^2$
	$= 15\,625$
Standard deviation	$= \sqrt{15\,625}$
	$= 125$

Exercise 8E

1 $\text{E}(X) = 4$, $\text{Var}(X) = 10$

Find

    a $\text{E}(2X)$,   **b** $\text{Var}(2X)$.

2 $\text{E}(X) = 2$, $\text{Var}(X) = 6$

Find

    a $\text{E}(3X)$,   **b** $\text{E}(3X + 1)$,   **c** $\text{E}(X - 1)$,   **d** $\text{E}(4 - 2X)$,

    e $\text{Var}(3X)$,   **f** $\text{Var}(3X + 1)$,   **g** $\text{Var}(X - 1)$.

3 The random variable X has a mean of 3 and a variance of 9.

Find

    a $\text{E}(2X + 1)$,   **b** $\text{E}(2 + X)$,   **c** $\text{Var}(2X + 1)$,   **d** $\text{Var}(2 + X)$.

4 The random variable X has a mean μ and standard deviation σ.

Find, in terms of μ and σ

    a $\text{E}(4X)$,   **b** $\text{E}(2X + 2)$,   **c** $\text{E}(2X - 2)$,

    d $\text{Var}(2X + 2)$,   **e** $\text{Var}(2X - 2)$.

5 The random variable Y has mean 2 and variance 9.

Find:

    a $\text{E}(3Y + 1)$,   **b** $\text{E}(2 - 3Y)$,   **c** $\text{Var}(3Y + 1)$,

    d $\text{Var}(2 - 3Y)$,   **e** $\text{E}(Y^2)$,   **f** $\text{E}[(Y - 1)(Y + 1)]$.

6 The random variable T has a mean of 20 and a standard deviation of 5.

It is required to scale T by using the transformation $S = 3T + 4$.

Find $\text{E}(S)$ and $\text{Var}(S)$.

7 A fair spinner is made from the disc in the diagram and the
random variable X represents the number it lands on after
being spun.

    a Write down the distribution of X.   **b** Work out $\text{E}(X)$.

    c Find $\text{Var}(X)$.   **d** Find $\text{E}(2X + 1)$.

    e Find $\text{Var}(3X - 1)$.

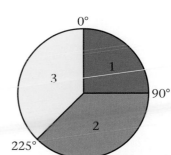

8 The discrete variable X has the probability distribution

x	-1	0	1	2
$P(X = x)$	0.2	0.5	0.2	0.1

a Find E(X),　　　　**b** Find Var (X),　　　　**c** Find E($\frac{1}{3}X + 1$),　　　　**d** Find Var($\frac{1}{3}X + 1$).

8.11 **Using a discrete uniform distribution as a model for the probability distribution of the outcomes of certain experiments.**

■ The probability distribution for the score S on a single roll of the die is:

s:	1	2	3	4	5	6
$P(S = s)$:	$\frac{1}{6}$	$\frac{1}{6}$	$\frac{1}{6}$	$\frac{1}{6}$	$\frac{1}{6}$	$\frac{1}{6}$

This is an example of a **discrete uniform distribution** over the set of values $\{1, 2, 3, 4, 5, 6\}$. It is called discrete because the values are discrete and it is called uniform because all the probabilities are the same.

■ Conditions for a discrete uniform distribution:
a **discrete random variable** X is defined over a set of n distinct values.

Each value is equally likely, i.e. $P(X = x) = \frac{1}{n}$ for each x.

■ In many cases X is defined over the set $\{1, 2, 3, ..., n\}$, in such cases the mean and variance are given by the following:

for a discrete uniform distribution X over the values 1, 2, 3, ..., n

$$E(X) = \frac{n + 1}{2}$$

$$Var(X) = \frac{(n + 1)(n - 1)}{12}$$

> Remember to use these formulae the values for X must be 1, 2, 3, ..., n.

You do not have to prove these results or know them for the examination, but you may find them useful in answering some questions.

Example 18

Digits are selected at random from a table of random numbers.

a Find the mean and standard deviation of a single digit.

b Find the probability that a particular digit lies within one standard deviation of the mean.

> Digits are single numbers. In this case the digits will be the numbers 0 to 9 inclusive.
> Let R represent this random variable having a discrete uniform distribution over the set $\{0, 1, 2, ..., 9\}$.
> Let X represent a random variable having a discrete uniform distribution over the set $\{1, 2, 3, ..., 10\}$.
> There is a simple relationship between X and R, namely $R = X - 1$.
> By introducing the random variable X you can use the standard formulae above for X with $n = 10$.

a $E(R) = E(X - 1)$

$\quad\quad = E(X) - 1$

$\quad\quad = \dfrac{n + 1}{2} - 1$ •—————————— Using the formula for the mean of a series 1, 2, ..., n.

$\quad\quad = \dfrac{10 + 1}{2} - 1$

$\quad\quad = 4.5$

and

$\quad\quad Var(R) = Var(X - 1)$

$\quad\quad\quad = Var(X)$

$\quad\quad\quad = \dfrac{(n + 1)(n - 1)}{12}$ •—————— Using the formula for the variance of a series 1, 2, ..., n.

$\quad\quad\quad = \dfrac{11 \times 9}{12}$

$\quad\quad\quad = 8.25$

Standard deviation

$\quad\quad \sigma = \sqrt{8.25}$

$\quad\quad\quad = 2.87 \ (3 \text{ s.f.})$

b Using the value of σ in **a**, the required probability is

$P(4.5 - 2.87... < R < 4.5 + 2.87...)$

$= P(1.63... < R < 7.37...)$

$= P(2 \leqslant R \leqslant 7)$

$= \dfrac{6}{10} = \dfrac{3}{5}$

Example 18 shows how you can use the formulae for the series $\{0, 1, 2 ..., n\}$ and the formulae for functions of X to get the answers. You could also get the answers by writing down the probability distribution and finding the mean and variance from scratch. As far as the examination is concerned this is a perfectly acceptable alternative but it may take a bit longer.

The discrete uniform distribution, despite its simplicity, arises in a number of instances. When using it as a model, care should be taken that all the conditions apply. One condition is that the probabilities are all the same. The other condition is that the variable is defined over a set of n distinct values. Here is an example. A bag contains five coins: one 50p, one 20p, one 10p, one 5p and one 2p. A child is told he can place his hand in the bag and keep the first coin he takes out. The random variable V represents the value of that coin. You could try to model V as a uniform distribution over the set $\{2, 5, 10, 20, 50\}$, but it is not clear that the probability of selecting the 50p coin is the same as that of the 10p coin, as they are of different sizes.

Exercise 8F

1 X is a discrete uniform distribution over the numbers 1, 2, 3, 4 and 5. Work out the expectation and variance of X.

2 Seven similar balls are placed in a bag. The balls have the numbers 1 to 7 on them. A ball is drawn out of the bag. The variable X represents the number on the ball.

  a Find $E(X)$.

  b Work out $Var(X)$.

3 A fair die is thrown once and the random variable X represents the value on the upper face.

  a Find the expectation and variance of X.

  b Calculate the probability that X is within one standard deviation of the mean.

4 A card is selected at random from a pack of cards containing the even numbers 2, 4, 6, ..., 20. The variable X represents the number on the card.

  a Find $P(X > 15)$.

  b Find the expectation and variance of X.

5 Repeat Question 4 for the odd numbers 1, 3, 5, ..., 19.

6 A straight line is drawn on a piece of paper. The line is divided into four equal lengths and the segments are marked 1, 2, 3 and 4. In a party game a person is blindfolded and asked to mark a point on the line and the number of the segment is recorded. A discrete uniform distribution over the set (1, 2, 3, 4) is suggested as model for this distribution. Comment on this suggestion.

7 In a fairground game the spinner shown is used.
It cost 5p to have a go on the spinner.

The spinner is spun and the number of pence shown is returned to the contestant.

If X is the number which comes up on the next spin,

  a name a suitable model for X,

  b find $E(X)$,

  c find $Var(X)$,

  d explain why you should not expect to make money at this game if you have a large number of goes.

Mixed exercise 8G

1 The random variable X has probability function

$$P(X = x) = \frac{x}{21} \qquad x = 1, 2, 3, 4, 5, 6.$$

a Construct a table giving the probability distribution of X.

Find

b $P(2 < X \leqslant 5)$,　　　　**c** $E(X)$,

d $Var(X)$,　　　　**e** $Var(3 - 2X)$.

2 The discrete random variable X has the probability distribution shown.

x	-2	-1	0	1	2	3
$P(X = x)$	0.1	0.2	0.3	r	0.1	0.1

Find

a r,　　　　**b** $P(-1 \leqslant X < 2)$,　　　　**c** $F(0.6)$,

d $E(2X + 3)$,　　　　**e** $Var(2X + 3)$.

3 A discrete random variable X has the probability distribution shown in the table below.

x	0	1	2
$P(X = x)$	$\frac{1}{5}$	b	$\frac{1}{5} + b$

a Find the value of b.　　　　**b** Show that $E(X) = 1.3$.

c Find the exact value of $Var(X)$.　　　　**d** Find the exact value of $P(X \leqslant 1.5)$.

4 The discrete random variable X has a probability function

$$P(X = x) = \begin{cases} k(1 - x) & x = 0, 1 \\ k(x - 1) & x = 2, 3 \\ 0 & \text{otherwise} \end{cases}$$

where k is a constant.

a Show that $k = \frac{1}{4}$.　　　　**b** Find $E(X)$ and show that $E(X^2) = 5.5$.

c Find $Var(2X - 2)$.

5 A discrete random variable X has the probability distribution,

x	0	1	2	3
$P(X = x)$	$\frac{1}{4}$	$\frac{1}{2}$	$\frac{1}{8}$	$\frac{1}{8}$

Find

a $P(1 < X \leqslant 2)$,　　　　**b** $F(1.5)$,　　　　**c** $E(X)$,

d $E(3X - 1)$,　　　　**e** $Var(X)$.

6 A discrete random variable is such that each of its values is assumed to be equally likely.

a Write the name of the distribution.   **b** Give an example of such a distribution.

A discrete random variable X as defined above can take values 0, 1, 2, 3, and 4.

Find

c E(X),   **d** Var (X).

7 The random variable X has a probability distribution

x	1	2	3	4	5
$P(X = x)$	0.1	p	q	0.3	0.1

a Given that E(X) = 3.1, write down two equations involving p and q.

Find

b the value of p and the value of q,

c Var(X),

d Var($2X - 3$).

8 The random variable X has probability function

$$P(X = x) = \begin{cases} kx & x = 1, 2 \\ k(x - 2) & x = 3, 4, 5 \end{cases}$$

where k is a constant.

a Find the value of k.

b Find the exact value of E(X).

c Show that, to three significant figures, Var(X) = 2.02.

d Find, to one decimal place, Var($3 - 2X$).

9 The random variable X has the discrete uniform distribution

$$P(X = x) = \frac{1}{6} \qquad x = 1, 2, 3, 4, 5, 6.$$

a Write down E(X) and show that Var(X) = $\frac{35}{12}$.

b Find E($2X - 1$).

c Find Var($3 - 2X$).

10 The random variable X has probability function

$$p(x) = \frac{(3x - 1)}{26} \qquad x = 1, 2, 3, 4.$$

a Construct a table giving the probability distribution of X.

Find

b $P(2 < X \leq 4)$,

c the exact value of E(X).

d Show that Var(X) = 0.92 to two significant figures.

e Find Var($1 - 3X$).

Summary of key points

1   For a random variable X
    - x is a particular value of X,
    - $P(X = x)$ or $p(x)$ refers to the probability that X is equal to a particular value x.

2   For a discrete random variable: the sum of all the probabilities must add up to one. or, in symbols, $\sum P(X = x) = 1$.

3   The cumulative frequency distribution $F(x) = P(X \leq x)$.

4   Expected value of X $E(X) = \sum x P(X = x) = \sum x p(x)$.

5   Variance of X $Var(X) = E(X^2) - [E(X)]^2$

6   $E(aX + b) = aE(X) + b$
    $Var(aX + b) = a^2\, Var(X)$

7   Conditions for a discrete uniform distribution:
    - A discrete random variable X is defined over a set of n distinct values.
    - Each value is equally likely.

8   For a discrete uniform distribution X over the values 1, 2, 3, …, n
    - $E(X) = \dfrac{n + 1}{2}$
    - $Var(X) = \dfrac{(n + 1)(n - 1)}{12}$

9

Normal distribution

After completing this chapter you should be able to

- use the normal distribution and its tables to find probabilities
- use the normal distribution and its tables to find means
- use the normal distribution and its tables to find standard deviations.

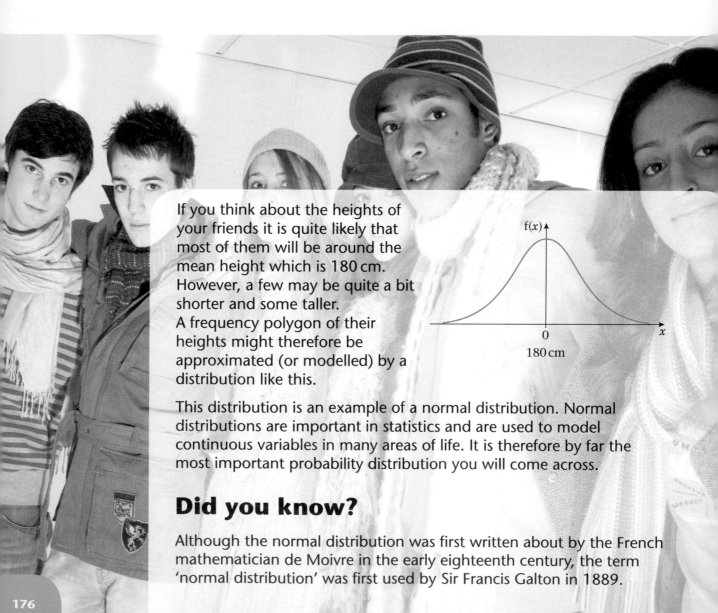

If you think about the heights of your friends it is quite likely that most of them will be around the mean height which is 180 cm. However, a few may be quite a bit shorter and some taller.
A frequency polygon of their heights might therefore be approximated (or modelled) by a distribution like this.

f(x)

0
180 cm

x

This distribution is an example of a normal distribution. Normal distributions are important in statistics and are used to model continuous variables in many areas of life. It is therefore by far the most important probability distribution you will come across.

Did you know?

Although the normal distribution was first written about by the French mathematician de Moivre in the early eighteenth century, the term 'normal distribution' was first used by Sir Francis Galton in 1889.

9.1 You can use tables to find probabilities of the standard normal distribution Z.

■ For a normal distribution probabilities are given by areas under the curve in a similar way to which frequencies are found by areas under a histogram or frequency polygon.

■ Tables are provided to help us calculate probabilities for the standard normal distribution, Z.

■ The standard normal variable is usually denoted by Z and has a mean of 0 and a standard deviation of 1. The usual way of writing this is

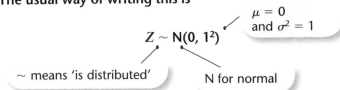

$$Z \sim N(0, 1^2)$$

$\mu = 0$ and $\sigma^2 = 1$

~ means 'is distributed'

N for normal

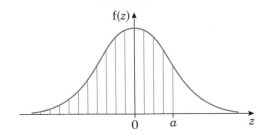

The mathematical equation of this curve is designed so that the total area under the curve $= 1$ (as the sum of the probabilities in a probability distribution always add up to 1) and so $P(Z \leqslant a)$ will just be the area under the curve to the left of a.

You can use tables or a calculator to find these probabilities. You will find a copy of the tables on pages 201 and 202.
The tables give $P(Z < z)$ for different values of z.
$\Phi(z)$ is often used as shorthand for $P(Z < z)$.

■ $P(Z > x) = 1 - P(Z < x)$ since the total area under the curve $= 1$.

■ For continuous distributions like the normal, there is no difference between $P(Z < z)$ and $P(Z \leqslant z)$.

Example 1

Use tables to find
a $P(Z < 1.52)$, **b** $P(Z > 2.60)$, **c** $P(Z < -0.75)$, **d** $P(-1.18 < Z < 1.43)$.

a

f(z)

−3 −2 −1 0 1 2 3 z
1.52

$P(Z < 1.52) = 0.9357$

Draw a diagram and shade the required region.

Notice that the required area is the area given in tables.

Always quote values from tables in full. You can round to 3 s.f. later if you feel this is appropriate.

b

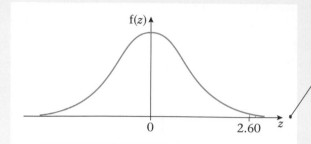

$P(Z > 2.60) = 1 - P(Z < 2.60)$

$= 1 - 0.9953$

$= 0.0047$

Draw a diagram and shade the required region.

The tables give $P(Z < 2.60)$ so you want $1 -$ this probability.

Because the normal distribution is describing a continuous variable there is no difference between $P(Z \leqslant 2.60)$ and $P(Z < 2.60)$.

c

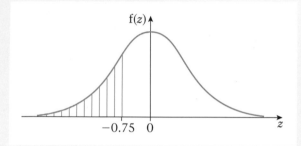

By symmetry $P(Z < -0.75) = P(Z > 0.75)$

$= 1 - 0.7734$

$= 0.2266$

Draw a diagram and shade the required region.

The tables do not give values of $z < 0$. So **use symmetry** to see that the required probability is the same as $P(Z > 0.75)$.

As in **b** use $1 - P(Z < 0.75)$.

d

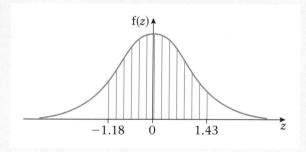

$P(-1.18 < Z < 1.43)$

$= P(-1.18 < Z < 0) + P(0 < Z < 1.43)$

$= (0.8810 - 0.5) + (0.9236 - 0.5)$

$= 0.3810 + 0.4236$

$= 0.8046$

Draw a diagram and shade the required area.

Think of a strategy to find the probabilities.

By symmetry $P(-1.18 < Z < 0) = P(0 < Z < 1.18)$.

Since area under curve $= 1$, by symmetry $P(Z < 0) = 0.5$. So $P(0 < Z < 1.18) = P(Z < 1.18) - 0.5$.

An alternative approach would be to say that $P(-1.18 < Z < 1.43) = P(Z < 1.43) - P(Z < -1.18)$ and then, using the approach from part **c** write $P(Z < -1.18)$ as $1 - P(Z < 1.18)$. You are advised to draw a diagram and select a suitable strategy.

Exercise 9A

Use tables of the normal distribution to find the following.

1   **a** $P(Z < 2.12)$ **b** $P(Z < 1.36)$

      c $P(Z > 0.84)$ **d** $P(Z < -0.38)$

2   **a** $P(Z > 1.25)$ **b** $P(Z > -1.68)$

      c $P(Z < -1.52)$ **d** $P(Z < 3.15)$

3   **a** $P(Z > -2.24)$ **b** $P(0 < Z < 1.42)$

      c $P(-2.30 < Z < 0)$ **d** $P(Z < -1.63)$

4   **a** $P(1.25 < Z < 2.16)$ **b** $P(-1.67 < Z < 2.38)$

      c $P(-2.16 < Z < -0.85)$ **d** $P(-1.57 < Z < 1.57)$

9.2   **You can use tables to find the value of _z_ given a probability.**

■   The table of **percentage points of the normal distribution** gives the value of _z_ for various values of $p = P(Z > z)$.

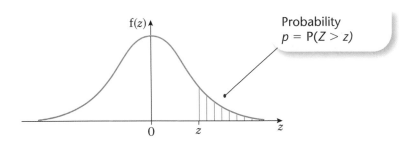

You should note that, by symmetry, if say $P(Z > 1.2816) = 0.1$ then $P(Z < -1.2816) = 0.1$ as well.

■   Whenever possible this table should be used to find _z_ given a value for $p = P(Z > z)$.

■   If $P(Z < a)$ is greater than 0.5, then _a_ will be > 0.
    If $P(Z < a)$ is less than 0.5, then _a_ is less than 0.

■   If $P(Z > a)$ is less than 0.5, then _a_ will be > 0.
    If $P(Z > a)$ is more than 0.5, then _a_ will be < 0.

Example 2

Find the value of the constant a such that

a $P(Z < a) = 0.7611$ **b** $P(Z > a) = 0.01$

c $P(Z > a) = 0.0287$ **d** $P(Z < a) = 0.0170$

a

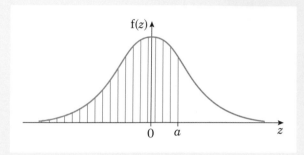

$P(Z < a) = 0.7611$

$a = 0.71$

Draw a diagram.

Look in the main table to find the value of z that gives $P(Z < z) = 0.7611$. This gives the value of a.

b

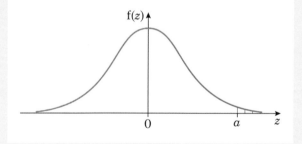

$P(Z > a) = 0.01$

$a = 2.3263$

Draw a diagram and note that $P(Z > a) < 0.5$ so a is positive.

Check the table of percentage points of the normal distribution to see if $p = 0.01$ is listed.

Using the table of percentage points.

c

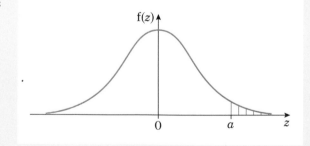

$P(Z > a) = 0.0287$

So $P(Z \leqslant a) = 1 - P(Z > a)$

$= 1 - 0.0287 = 0.9713$

$a = 1.90$

$P(Z > a) < 0.5$ so a is > 0 but $p = 0.0287$ is not listed in the table of percentage points of the normal distribution.

Since $p = 0.0287$ is not in the tables, calculate $1 - 0.0287$ and use the main table. Find z so that $P(Z < z) = 0.9713$.

d

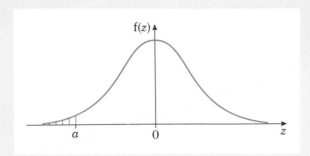

Since $P(Z < a) < 0.5$, then $a < 0$.

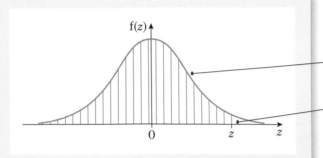

Use Symmetry.

This will be $1 - 0.0170 = 0.9830$.

This is $p = 0.0170$.

$1 - P(Z < a) = 1 - 0.0170 = 0.9830$

$P(Z < z) = 0.9830$

Implies $z = 2.12$

Therefore $a = -2.12$

Since $p = 0.0170$ is not in the tables, use the main table to find the value of z so that $P(Z < z) = 0.9830$.

Then a is simply $-z$ or -2.12.

Exercise 9B

Find the value of a in the following

1   **a** $P(Z < a) = 0.9082$ **b** $P(Z > a) = 0.0314$
     c $P(Z < a) = 0.3372$ **d** $P(Z > a) = 0.6879$

2   **a** $P(Z < a) = 0.9938$ **b** $P(Z > a) = 0.4129$
     c $P(Z > a) = 0.7611$ **d** $P(Z > a) = 0.2000$

3   **a** $P(Z > a) = 0.1500$ **b** $P(Z > a) = 0.9500$
     c $P(Z > a) = 0.1112$ **d** $P(Z < a) = 0.9990$

4   **a** $P(0 < Z < a) = 0.3554$ **b** $P(0 < Z < a) = 0.4946$
     c $P(-a < Z < a) = 0.5820$ **d** $P(-a < Z < a) = 0.8230$

5   **a** $P(0 < Z < a) = 0.10$ **b** $P(0 < Z < a) = 0.35$
     c $P(-a < Z < a) = 0.80$ **d** $P(-a < Z < a) = 0.40$

■ A normal distribution can be used to model a number of different variables.

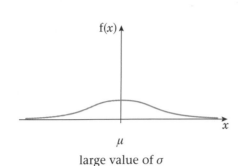

large value of σ

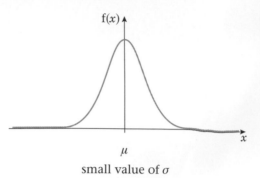

small value of σ

> The distribution will have the same familiar 'bell' shape but the mean μ and the standard deviation σ will be different.

■ The random variable X having a mean of μ and a standard deviation of σ (or variance σ^2) can be written as $X \sim N(\mu, \sigma^2)$.

■ $X \sim N(\mu, \sigma^2)$ can be transformed into the random variable $Z \sim N(0, 1^2)$ by the formula

$$Z = \frac{X - \mu}{\sigma}.$$

■ You can round your *z* value to 2 s.f. to use the nearest value in the tables.

Example 3

The random variable $X \sim N(50, 4^2)$.

Find

a $P(X < 53)$,   **b** $P(X \leqslant 45)$.

> $X \sim N(50, 4^2)$ means that the $\mu = 50$ and $\sigma = 4$.

Method 1

a $P(X < 53) = P\left(Z < \dfrac{53 - 50}{4}\right)$ Use $Z = \dfrac{X - \mu}{\sigma}$.

$\phantom{P(X < 53)}= P(Z < 0.75)$

$\phantom{P(X < 53)}= 0.7734$ Using the table with $z = 0.75$.

Method 2

Draw two diagrams showing 50, 53 for X and their equivalent values for Z.

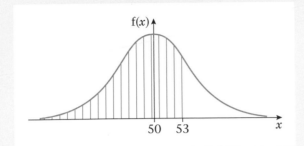
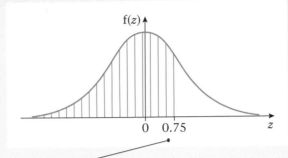

This value of 0.75 is found using $Z = \dfrac{53 - 50}{4}$

The probability $P(Z < 0.75) = 0.7734$ is found from tables.

b $P(X \leqslant 45) = P\left(Z \leqslant \dfrac{45 - 50}{4}\right)$

Use $Z = \dfrac{X - \mu}{\sigma}$

$ = P(Z < -1.25)$

$ = P(Z > 1.25)$

Remember that when using the normal distribution there is no difference between $Z \leqslant a$ and $Z < a$.

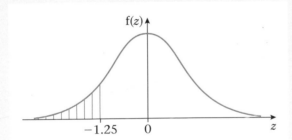

Draw a diagram and use symmetry.
$P(Z < -1.25) = P(Z > 1.25)$.

$ = 1 - P(Z < 1.25)$

$ = 1 - 0.8944$

$ = 0.1056$

Remember the main table gives $P(Z < z)$ so you need to use $1 - P(Z < z)$ to calculate $P(Z > z)$.

Use of calculators

You may be able to find probabilities for a normal distribution using your calculator. You will need to refer to the instructions for your particular calculator, but some give $P(a < X < b)$ if you enter: normalcdf(a, b, μ, σ). In the above example to find $P(X < 53)$ you could enter normalcdf(50, 53, 50, 4) (which would give you $P(50 < X < 53)$) and then add 0.5 (to give $P(X < 50)$), whereas to find $P(X < 45)$ you could enter normalcdf(45, 50, 50, 4) and subtract your answer from 0.5.

These calculators can be used in your S1 examination but you are advised to state clearly the probability you are finding and give your final answer to 3 s.f.
E.g. $P(X < 53) = 0.5 + 0.27337\ldots = 0.77337\ldots = 0.773$.

Example 4

The random variable $Y \sim N(20, 9)$.

Find the value of b such that $P(Y > b) = 0.0668$.

Method 1

$Y \sim N(20, 9)$ means that $\mu = 20$ and $\sigma = 3$
(the variance is 9 so standard deviation is 3).

$$P(Y > b) = 0.0668$$

Use $Z = \dfrac{Y - \mu}{\sigma}$

$$P\left(Z > \frac{b - 20}{3}\right) = 0.0668$$

$p = 0.0668$ is not in the table of percentage points.

So $P\left(Z < \dfrac{b - 20}{3}\right) = 1 - 0.0668$

$$= 0.9332$$

Calculate $1 - 0.0668$ and use the main table.

So $\dfrac{b - 20}{3} = 1.50$

$$b - 20 = 4.50$$

$$b = 24.50$$

$P(Z < 1.50) = 0.9332$ so use $z = 1.50$ to form an equation in b and solve.

Method 2

Draw two diagrams one for Y showing 20, b and probability and a parallel one for Z.

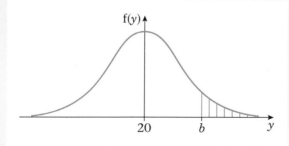

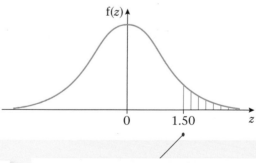

$$1.50 = \frac{b - 20}{3}$$

$$b = 3 \times 1.50 + 20$$

$$b = 24.50$$

Use $Z = \dfrac{Y - \mu}{\sigma}$ to link Y and Z values and obtain an equation in b.

1.50 comes from tables of $P(Z < z)$ using $1 - 0.0668 = 0.9332$ and $P(Z < 1.50) = 0.9332$.

Exercise 9C

1 The random variable $X \sim N(30, 2^2)$.

Find   **a** $P(X < 33)$,   **b** $P(X > 26)$.

2 The random variable $X \sim N(40, 9)$.

Find **a** $P(X > 45)$, **b** $P(X < 38)$.

3 The random variable $Y \sim N(25, 25)$.

Find **a** $P(Y < 20)$, **b** $P(18 < Y < 26)$.

4 The random variable $X \sim N(18, 10)$.

Find **a** $P(X > 20)$, **b** $P(X < 15)$.

5 The random variable $X \sim N(20, 8)$.

Find **a** $P(X > 15)$, **b** the value of a such that $P(X < a) = 0.8051$.

6 The random variable $Y \sim N(30, 5^2)$.

Find the value of a such that $P(Y > a) = 0.30$.

7 The random variable $X \sim N(15, 3^2)$.

Find the value of a such that $P(X > a) = 0.15$.

8 The random variable $X \sim N(20, 12)$.

Find the value of a and the value of b such that

a $P(X < a) = 0.40$, **b** $P(X > b) = 0.6915$.

c Write down $P(b < X < a)$.

9 The random variable $Y \sim N(100, 15^2)$.

Find the value of a and the value of b such that

a $P(Y > a) = 0.975$, **b** $P(Y < b) = 0.10$.

c Write down $P(a < Y < b)$.

10 The random variable $X \sim N(80, 16)$.

Find the value of a and the value of b such that

a $P(X > a) = 0.40$, **b** $P(X < b) = 0.5636$.

c Write down $P(b < X < a)$.

9.4 You can use normal tables to find μ and σ.

■ If $X \sim N(\mu, \sigma^2)$ and $P(X > a) = \alpha$, where α is a probability, you write this statement as

$$P\left(Z > \frac{a - \mu}{\sigma}\right) = \alpha.$$

■ Sometimes neither μ nor σ is given, in which case you will have to solve simultaneous equations.

There are four unknowns: a, μ, σ, α and in most questions in S1 you will be given three of these values and asked to find the fourth.

In Example 3 of 9.3 you saw how to find α, if a, μ, and σ, are given and in Example 4 you saw how to find a if you are given μ, σ and α.

A similar method can be used to find μ or σ.

Example 5

The random variable $X \sim N(\mu, 3^2)$.

Given that $P(X > 20) = 0.20$, find the value of μ.

$P(X > 20) = 0.20$

$P\left(Z > \dfrac{20 - \mu}{3}\right) = 0.20$

Use $Z = \dfrac{X - \mu}{\sigma}$.

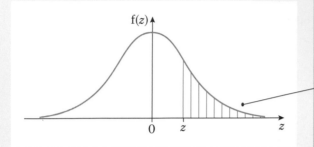

Draw a diagram.

$p = 0.20$

Use the table of percentage points of the normal distribution with $p = 0.20$ to get $z = 0.8416$.

From tables $P(Z > 0.8416) = 0.20$

So $0.8416 = \dfrac{20 - \mu}{3}$

So $\mu = 20 - 3 \times 0.8416$

$\mu = 17.4752$ or 17.5 (3 s.f.)

Use $z = 0.8416$ to form an equation in μ and solve.

Example 6

The random variable $X \sim N(50, \sigma^2)$.

Given that $P(X < 46) = 0.2119$, find the value of σ.

$P(X < 46) = 0.2119$

$P\left(Z < \dfrac{46 - 50}{\sigma}\right) = 0.2119$

Use $Z = \dfrac{X - \mu}{\sigma}$.

Prob. = 0.2119

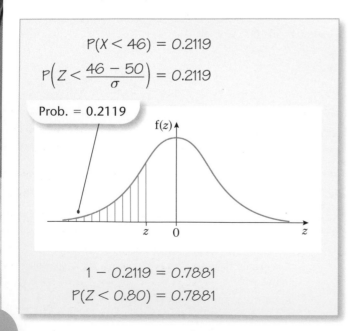

Draw a diagram for Z.

Since $P(Z < z) < 0.5$, use symmetry and calculate $1 - 0.2119 = 0.7881$.

$1 - 0.2119 = 0.7881$

$P(Z < 0.80) = 0.7881$

From tables, find the value which corresponds to 0.7881.

So $\qquad \dfrac{46 - 50}{\sigma} = -0.80$

N.B. the minus sign since z is to the left of the mean 0.

$\dfrac{-4}{-0.80} = \sigma$

Form an equation in σ and solve.

So $\qquad \sigma = 5$

N.B. σ must always be > 0. So if your answer for σ is negative, check to see if a minus sign has been missed out.

Example 7

The random variable $X \sim N(\mu, \sigma^2)$.

Given that $P(X > 35) = 0.025$ and $P(X < 15) = 0.1469$, find the value of μ and the value of σ.

Draw a diagram for X and a diagram for Z.
Note that $35 > \mu$ and $15 < \mu$.

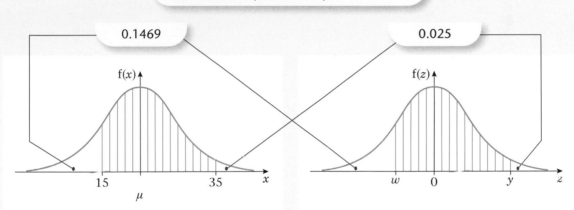

The table of percentage points of the normal distribution give:

$P(Z > 1.96) = 0.025$

So $\qquad y = 1.96$

You will need to use tables to find w and y.

Using $p = 0.025$ gives $z = 1.96$ from the tables.

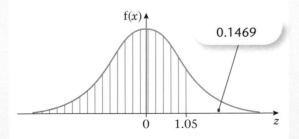

To find w you will need to use symmetry and calculate $1 - 0.1469$. This gives value of Z as 1.05.

From the main table find
$P(Z < 1.05) = 0.8531 \ (= 1 - 0.1469)$.

Since w is below the mean

$w = -1.05$

N.B. minus sign since the value of z is to the left of the mean $= 0$.

So $\qquad -1.05 = \dfrac{15 - \mu}{\sigma}$

and $\quad 1.96 = \dfrac{35 - \mu}{\sigma}$

> Use $\dfrac{X - \mu}{\sigma}$ to link X and Z values and form two equations in μ and σ.

i.e. $\quad -1.05\sigma = 15 - \mu$

and $\quad 1.96\sigma = 35 - \mu$

> Multiply by σ.

$\qquad 3.01\sigma = 20$

> Subtract to solve for σ.

So $\qquad \sigma = 6.6445\ldots$

$\qquad \mu = 35 - 1.96\sigma$

> Substitute back in one of the equations and solve for μ.

So $\qquad \mu = 21.9767$

Therefore $\mu = 22.0$ and $\sigma = 6.64$ (to 3 s.f.)

Exercise 9D

1. The random variable $X \sim N(\mu, 5^2)$ and $P(X < 18) = 0.9032$.
   Find the value of μ.

2. The random variable $X \sim N(11, \sigma^2)$ and $P(X > 20) = 0.01$.
   Find the value of σ.

3. The random variable $Y \sim N(\mu, 40)$ and $P(Y < 25) = 0.15$.
   Find the value of μ.

4. The random variable $Y \sim N(50, \sigma^2)$ and $P(Y > 40) = 0.6554$.
   Find the value of σ.

5. The random variable $X \sim N(\mu, \sigma^2)$.
   Given that $P(X < 17) = 0.8159$ and $P(X < 25) = 0.9970$, find the value of μ and the value of σ.

6. The random variable $Y \sim N(\mu, \sigma^2)$.
   Given that $P(Y < 25) = 0.10$ and $P(Y > 35) = 0.005$, find the value of μ and the value of σ.

7. The random variable $X \sim N(\mu, \sigma^2)$.
   Given that $P(X > 15) = 0.20$ and $P(X < 9) = 0.20$, find the value of μ and the value of σ.

   > **Hint** for Question 7:
   > Draw a diagram and use symmetry to find μ.

8. The random variable $X \sim N(\mu, \sigma^2)$.
   The lower quartile of X is 25 and the upper quartile of X is 45.
   Find the value of μ and the value of σ.

9. The random variable $X \sim N(0, \sigma^2)$.
   Given that $P(-4 < X < 4) = 0.6$, find the value of σ.

10. The random variable $X \sim N(2.68, \sigma^2)$.
    Given that $P(X > 2a) = 0.2$ and $P(X < a) = 0.4$, find the value of σ and the value of a.

9.5 You can use the normal distribution to answer questions in context.

Example 8

The heights of a large group of women are normally distributed with a mean of 165 cm and a standard deviation of 3.5 cm. A woman is selected at random from this group.

a Find the probability that she is shorter than 160 cm.

Steven is looking for a woman whose height is between 168 cm and 174 cm for a part in his next film.

b Find the proportion of women from this group who meet Steven's criteria.

Let H = the height of a woman from this group

$H \sim N(165, 3.5^2)$

a $P(H < 160) = P\left(Z < \dfrac{160 - 165}{3.5}\right)$

Use $Z = \dfrac{H - \mu}{\sigma}$.

$= P(Z < -1.4285\ldots)$

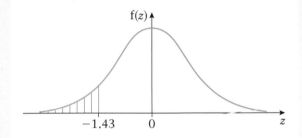

You should round Z to the nearest value in the tables.

You are not required to interpolate.

$P(Z < -1.43) = 1 - P(Z < 1.43)$

$= 1 - 0.9236$

$= 0.0764$

Use symmetry.

If you used a calculator to find $P(Z < -1.4285\ldots)$ you would obtain an answer of 0.07656... In the S1 examination the examiners would accept answers in the range 0.0764 to 0.0766.

b You require $P(168 < H < 174)$

$= P\left(\dfrac{168 - 165}{3.5} < Z < \dfrac{174 - 165}{3.5}\right)$

Using $Z = \dfrac{H - \mu}{\sigma}$ twice.

$= P(0.857\ldots < Z < 2.571\ldots)$

$= P(0.86 < Z < 2.55)$

$= P(Z < 2.55) - P(Z < 0.86)$

Round each z value to the nearest value in the tables. Note that 2.571 is nearer 2.55 than 2.60.

$= 0.9946 - 0.8051$

$= 0.1895$

If you use a calculator to find this probability you could obtain 0.19061... for the probability. In a case like this the examiners in S1 would accept any answer which rounds to 0.19.

Example **9**

Boxes of chocolates are produced with a mean weight of 510 g. Quality control checks show that 1% of boxes are rejected because their weight is less than 485 g.

a Find the standard deviation of the weight of a box of chocolates.

b Hence find the proportion of boxes that weigh more than 525 g.

a Let W = weight of a box of chocolates.

$$W \sim N(510, \sigma^2)$$

$$P(W < 485) = 0.01$$

$$P\left(Z < \frac{485 - 510}{\sigma}\right) = 0.01$$

So $\quad \dfrac{485 - 510}{\sigma} = -2.3263$

$$\sigma = 10.746....$$

$$\sigma = 10.7 \ (3 \ \text{s.f.})$$

> Write the information in the question in terms of a probability statement.

> Use $Z = \dfrac{W - \mu}{\sigma}$.

> Using the table of percentage points of the normal distribution with $p = 0.01$.
> Notice the minus sign, since you have $P(Z < z) = 0.01$.

b $\quad P(W > 525) = P\left(Z > \dfrac{525 - 510}{10.7...}\right)$

$$= P(Z > 1.40)$$

$$= 1 - 0.9192$$

$$= 0.0808$$

> Round z to 2 d.p. to use the tables.

> Remember to use $P(Z > z) = 1 - P(Z < z)$.

Once again a range of different answers can be obtained using more figures for σ or using a calculator to find the probability. Typically the examiners would accept any answer rounding to 0.081 or 0.080.

Mixed exercise **9E**

1 The heights of a large group of men are normally distributed with a mean of 178 cm and a standard deviation of 4 cm.

A man is selected at random from this group.

a Find the probability that he is taller than 185 cm.

A manufacturer of door frames wants to ensure that fewer than 0.005 men have to stoop to pass through the frame.

b On the basis of this group, find the minimum height of a door frame.

2 The weights of steel sheets produced by a factory are known to be normally distributed with mean 32.5 kg and standard deviation 2.2 kg.

   a Find the percentage of sheets that weigh less than 30 kg.

  Bob requires sheets that weigh between 31.6 kg and 34.8 kg.

   b Find the percentage of sheets produced that satisfy Bob's requirements.

3 The time a mobile phone battery lasts before needing to be recharged is assumed to be normally distributed with a mean of 48 hours and a standard deviation of 8 hours.

   a Find the probability that a battery will last for more than 60 hours.

   b Find the probability that the battery lasts less than 35 hours.

4 The random variable $X \sim N(24, \sigma^2)$.

  Given that $P(X > 30) = 0.05$, find

   a the value of σ,

   b $P(X < 20)$,

   c the value of d so that $P(X > d) = 0.01$.

5 A machine dispenses liquid into plastic cups in such a way that the volume of liquid dispensed is normally distributed with a mean of 120 ml. The cups have a capacity of 140 ml and the probability that the machine dispenses too much liquid so that the cup overflows is 0.01.

   a Find the standard deviation of the volume of liquid dispensed.

   b Find the probability that the machine dispenses less than 110 ml.

  Ten percent of customers complain that the machine has not dispensed enough liquid.

   c Find the largest volume of liquid that will lead to a complaint.

6 The random variable $X \sim N(\mu, \sigma^2)$. The lower quartile of X is 20 and the upper quartile is 40.

  Find μ and σ.

7 The heights of seedlings are normally distributed. Given that 10% of the seedlings are taller than 15 cm and 5% are shorter than 4 cm, find the mean and standard deviation of the heights.

8 A psychologist gives a student two different tests. The first test has a mean of 80 and a standard deviation of 10 and the student scored 85.

   a Find the probability of scoring 85 or more on the first test.

  The second test has a mean of 100 and a standard deviation of 15. The student scored 105 on the second test.

   b Find the probability of a score of 105 or more on the second test.

   c State, giving a reason, which of the student's two test scores was better.

9 Jam is sold in jars and the mean weight of the contents is 108 grams. Only 3% of jars have contents weighing less than 100 grams. Assuming that the weight of jam in a jar is normally distributed find

a the standard deviation of the weight of jam in a jar,

b the proportion of jars where the contents weigh more than 115 grams.

10 The waiting time at a doctor's surgery is assumed to be normally distributed with standard deviation of 3.8 minutes. Given that the probability of waiting more than 15 minutes is 0.0446, find

a the mean waiting time,

b the probability of waiting fewer than 5 minutes.

11 The thickness of some plastic shelving produced by a factory is normally distributed. As part of the production process the shelving is tested with two gauges. The first gauge is 7 mm thick and 98.61% of the shelving passes through this gauge. The second gauge is 5.2 mm thick and only 1.02% of the shelves pass through this gauge.

Find the mean and standard deviation of the thickness of the shelving.

12 The random variable $X \sim N(14, 9)$. Find

    a $P(X \geqslant 11)$,   **b** $P(9 < X < 11)$.

13 The random variable $X \sim N(20, 5^2)$. Find

    a $P(X \leqslant 16)$,   **b** the value of d such that $P(X < d) = 0.95$.

Summary of key points

1 The random variable X that has a normal distribution with mean μ and standard deviation σ is represented by

$$X \sim N(\mu, \sigma^2)$$

where σ^2 is the variance of the normal distribution.

2 If $X \sim N(\mu, \sigma^2)$ and $Z \sim N(0, 1^2)$ then

$$Z = \frac{X - \mu}{\sigma}$$

N.B. Tables of Z are given in the formula book.

Review Exercise

1 As part of a statistics project, Gill collected data relating to the length of time, to the nearest minute, spent by shoppers in a supermarket and the amount of money they spent. Her data for a random sample of 10 shoppers are summarised in the table below, where t represents time and £m the amount spent over £20.

t (minutes)	£m
15	-3
23	17
5	-19
16	4
30	12
6	-9
32	27
23	6
35	20
27	6

a Write down the actual amount spent by the shopper who was in the supermarket for 15 minutes.

b Calculate S_{tt}, S_{mm} and S_{tm}.
(You may use $\sum t^2 = 5478$, $\sum m^2 = 2101$, and $\sum tm = 2485$)

c Calculate the value of the product moment correlation coefficient between t and m.

d Write down the value of the product moment correlation coefficient between t and the actual amount spent. Give a reason to justify your value.

On another day Gill collected similar data. For these data the product moment correlation coefficient was 0.178

e Give an interpretation to both of these coefficients.

f Suggest a practical reason why these two values are so different. **E**

2 The random variable X has probability function
$$P(X = x) = \frac{(2x - 1)}{36} \qquad x = 1, 2, 3, 4, 5, 6.$$

a Construct a table giving the probability distribution of X.

Find

b $P(2 < X \leqslant 5)$,

c the exact value of E(X).

d Show that Var(X) = 1.97 to three significant figures.

e Find Var(2 − 3X).

3 The measure of intelligence, IQ, of a group of students is assumed to be Normally distributed with mean 100 and standard deviation 15.

a Find the probability that a student selected at random has an IQ less than 91.

The probability that a randomly selected student as an IQ of at least 100 + k is 0.2090.

b Find, to the nearest integer, the value of k.

4 The scatter diagrams below were drawn by a student.

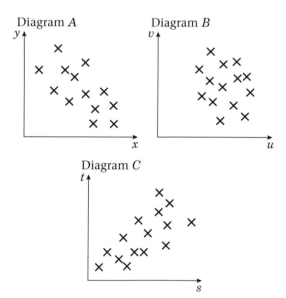

Diagram A

Diagram B

Diagram C

The student calculated the value of the product moment correlation coefficient for each of the sets of data.

The values were

0.68   −0.79   0.08

Write down, with a reason, which value corresponds to which scatter diagram. **E**

5 A long distance lorry driver recorded the distance travelled, m miles, and the amount of fuel used, f litres, each day. Summarised below are data from the driver's records for a random sample of eight days.

The data are coded such that $x = m − 250$ and $y = f − 100$.

$$\Sigma x = 130 \qquad \Sigma y = 48$$
$$\Sigma xy = 8880 \qquad S_{xx} = 20\,487.5$$

a Find the equation of the regression line of y on x in the form $y = a + bx$.

b Hence find the equation of the regression line of f on m.

c Predict the amount of fuel used on a journey of 235 miles. **E**

6 The random variable X has probability function

$$P(X = x) = \begin{cases} kx & x = 1, 2, 3, \\ k(x + 1) & x = 4, 5 \end{cases}$$

where k is a constant.

a Find the value of k.

b Find the exact value of E(X).

c Show that, to three significant figures, Var(X) = 1.47.

d Find, to one decimal place, Var(4 − 3X).

7 A scientist found that the time taken, M minutes, to carry out an experiment can be modelled by a normal random variable with mean 155 minutes and standard deviation 3.5 minutes.

Find

a P(M > 160).

b P(150 ⩽ M ⩽ 157).

c the value of m, to one decimal place, such that P(M ⩽ m) = 0.30.

8 The random variable X has probability distribution

x	1	2	3	4	5
$P(X = x)$	0.1	p	0.20	q	0.30

a Given that $E(X) = 3.5$, write down two equations involving p and q.

Find

b the value of p and the value of q,

c Var (X),

d Var $(3 - 2X)$.

9 A manufacturer stores drums of chemicals. During storage, evaporation takes place. A random sample of 10 drums was taken and the time in storage, x weeks, and the evaporation loss, y ml, are shown in the table below.

x	3	5	6	8	10	12	13	15	16	18
y	36	50	53	61	69	79	82	90	88	96

a On graph paper, draw a scatter diagram to represent these data.

b Give a reason to support fitting a regression model of the form $y = a + bx$ to these data.

c Find, to two decimal places, the value of a and the value of b.

(You may use $\sum x^2 = 1352$, $\sum y^2 = 53\,112$ and $\sum xy = 8354$.)

d Give an interpretation of the value of b.

e Using your model, predict the amount of evaporation that would take place after

i 19 weeks,

ii 35 weeks.

f Comment, with a reason, on the reliability of each of your predictions.

10 a Write down two reasons for using statistical models.

b Give an example of a random variable that could be modelled by

i a normal distribution,

ii a discrete uniform distribution.

11 The heights of a group of athletes are modelled by a normal distribution with mean 180 cm and standard deviation 5.2 cm. The weights of this group of athletes are modelled by a normal distribution with mean 85 kg and standard deviation 7.1 kg.

Find the probability that a randomly chosen athlete,

a is taller than 188 cm,

b weighs less than 97 kg.

c Assuming that for these athletes height and weight are independent, find the probability that a randomly chosen athlete is taller than 188 cm and weighs more than 97 kg.

d Comment on the assumption that height and weight are independent. **E**

12 A metallurgist measured the length, l mm, of a copper rod at various temperatures, t °C, and recorded the following results.

t	l
20.4	2461.12
27.3	2461.41
32.1	2461.73
39.0	2461.88
42.9	2462.03
49.7	2462.37
58.3	2462.69
67.4	2463.05

The results were then coded such that $x = t$ and $y = l - 2460.00$.

a Calculate S_{xy} and S_{xx}.

(You may use $\sum x^2 = 15\,965.01$ and $\sum xy = 757.467$)

b Find the equation of the regression line of y on x in the form $y = a + bx$.

c Estimate the length of the rod at 40°C.

d Find the equation of the regression line of l on t.

e Estimate the length of the rod at 90°C.

f Comment on the reliability of your estimate in **e**.   **E**

13 The random variable X has the discrete uniform distribution

$$P(X = x) = \tfrac{1}{5}, \qquad x = 1, 2, 3, 4, 5.$$

a Write down the value of $E(X)$ and show that $Var(X) = 2$.

Find

b $E(3X - 2)$,

c $Var(4 - 3X)$.   **E**

14 From experience a high jumper knows that he can clear a height of at least 1.78 m once in five attempts. He also knows that he can clear a height of at least 1.65 m on seven out of 10 attempts.

Assuming that the heights the high jumper can reach follow a Normal distribution,

a draw a sketch to illustrate the above information,

b find, to three decimal places, the mean and the standard deviation of the heights the high jumper can reach,

c calculate the probability that he can jump at least 1.74 m.   **E**

15 A young family were looking for a new three bedroom semi-detached house.

A local survey recorded the price x, in £1000s, and the distance y, in miles, from the station, of such houses. The following summary statistics were provided

$S_{xx} = 113\,573$, $S_{yy} = 8.657$, $S_{xy} = -808.917$

a Use these values to calculate the product moment correlation coefficient.

b Give an interpretation of your answer to **a**.

Another family asked for the distances to be measured in km rather than miles.

c State the value of the product moment correlation coefficient in this case.   **E**

16 A student is investigating the relationship between the price (y pence) of 100 g of chocolate and the percentage (x%) of cocoa solids in the chocolate.

The following data are obtained

Chocolate brand	x (% cocoa)	y (pence)
A	10	35
B	20	55
C	30	40
D	35	100
E	40	60
F	50	90
G	60	110
H	70	130

(You may use: $\sum x = 315$, $\sum x^2 = 15\,225$, $\sum y = 620$, $\sum y^2 = 56\,550$, $\sum xy = 28\,750$)

a Draw a scatter diagram to represent these data.

b Show that $S_{xy} = 4337.5$ and find S_{xx}.

The student believes that a linear relationship of the form $y = a + bx$ could be used to describe these data.

c Use linear regression to find the value of a and the value of b, giving your answers to one decimal place.

d Draw the regression line on your diagram.

The student believes that one brand of chocolate is overpriced.

e Use the scatter diagram to
  i state which brand is overpriced,
  ii suggest a fair price for this brand.
  Give reasons for both your answers.

17 The random variable X has a normal distribution with mean 20 and standard deviation 4.

a Find $P(X > 25)$.

b Find the value of d such that
$P(20 < X < d) = 0.4641$.

18 The random variable X has probability distribution

x	1	3	5	7	9
$P(X = x)$	0.2	p	0.2	q	0.15

a Given that $E(X) = 4.5$, write down two equations involving p and q.

Find

b the value of p and the value of q,

c $P(4 < X \leqslant 7)$.

Given that $E(X^2) = 27.4$, find

d $Var(X)$,

e $E(19 - 4X)$,

f $Var(19 - 4X)$.

19 The box plot in Figure 1 shows a summary of the weights of the luggage, in kg, for each musician in an orchestra on an overseas tour.

The airline's recommended weight limit for each musicians' luggage was 45 kg.

Given that none of the musician's luggage weighed exactly 45 kg,

a state the proportion of the musicians whose luggage was below the recommended weight limit.

A quarter of the musicians had to pay a charge for taking heavy luggage.

b State the smallest weight for which the charge was made.

c Explain what you understand by the ✗ on the box plot in Figure 1, and suggest an instrument that the owner of this luggage might play.

d Describe the skewness of this distribution. Give a reason for your answer.

One musician in the orchestra suggests that the weights of the luggage, in kg, can be modelled by a normal distribution with quartiles as given in Figure 1.

e Find the standard deviation of this normal distribution.

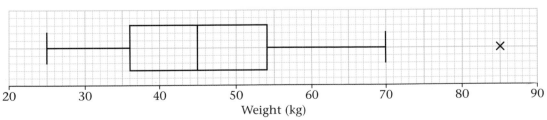

Figure 1

Examination style paper

1 A fair die has six faces numbered 1, 1, 1, 2, 2 and 3. The die is rolled twice and the number showing on the uppermost face is recorded.

Find the probability that the sum of the two numbers is at least three. (5)

2 Jars are filled with jam by a machine. Each jar states it contains 450 g of jam.

The actual weight of jam in each jar is normally distributed with mean 460 g and standard deviation of 10 g.

a Find the probability that a jar contains less then the stated weight of 450 g. (3)

Jars are sold in boxes of 30.

b Find the expected number of jars containing less than the stated weight. (2)

The mean weight is changed so that 1% of the jars contain less than the stated weight of 450 g of jam.

c Find the new mean weight of jam. (4)

3 A discrete random variable X has a probability distribution as shown in the table below

x	0	1	2	3
P($X = x$)	0.2	0.3	b	$2a$

where a and b are constants.

If E(X) = 1.6,

a show that $b = 0.2$ and find the value of a. (5)

Find

b E($5 - 2X$), (2)

c Var(X), (3)

d Var($5 - 2X$). (2)

4 The attendance at college of a group of 20 students was recorded for a four week period. The numbers of students attending each of the 16 classes are shown below.

20	20	19	19
18	19	18	20
20	16	19	20
17	19	20	18

a Calculate the mean and the standard deviation of the attendance. (4)

b Express the mean as a percentage of the 20 students in the group. (1)

In the same four week period, the attendance of a different group of 22 students was recorded.

22	18	20	21
17	16	16	17
20	17	18	19
18	20	17	16

c Find the mode, median and inter-quartile range for this group of students. (3)

A box plot for the first group of students is drawn below.

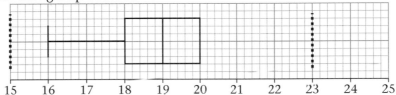

d Using the same scale draw a box plot for the second group of students. (3)

The mean percentage attendance and standard deviation for the second group of students are 82.95 and 1.82 respectively.

e Compare and contrast the attendance of each group of students. (3)

5 The random variable X has the distribution

$$P(X = x) = \frac{1}{n} \text{ for } x = 1, 2, \ldots, n.$$

a Write down the name of the distribution. (1)

Given that $E(X) = 10$,

b show that $n = 19$, (2)

c Find $Var(X)$. (2)

6 A researcher thinks that there is a link between a person's confidence and their height. She devises a test to measure the confidence, c, of nine people and their height, h cm. The data are shown in the table below.

h	179	169	187	166	162	193	161	177	168
c	569	561	579	561	540	598	542	565	573

$[\sum h = 1562,\ \sum c = 5088,\ \sum hc = 884\,484,\ \sum h^2 = 272\,094,\ \sum c^2 = 2\,878\,966]$

a Draw a scatter diagram to represent these data. (2)

b Find the values of S_{hh}, S_{cc} and S_{hc}. (3)

c Calculate the value of the product moment correlation coefficient. (3)

d Calculate the equation of the regression line of c on h. (4)

e Draw this line on your scatter diagram. (2)

f Interpret the gradient of the regression line. (1)

The researcher decides to use this regression model to predict a person's confidence.

g Find the proposed confidence for the person who has a height of 172 cm. (2)

7 A fairground game involves trying to hit a moving target with an air rifle pellet.
Each player has up to three pellets in a round. Five points are scored if a pellet hits the target, but the round is over if a pellet misses the target.

Jean has a constant probability of 0.4 of hitting the target.

The random variable X is the number of points Jean scores in a round.

Find

a the probability that Jean scores 15 points in a round, (2)

b the probability distribution of X. (5)

A game consists of two rounds.

c Find the probability that Jean scores more points in her second round than her first. (6)

Appendix

The normal distribution function

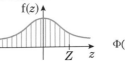

$$\Phi(z) = P(Z < z)$$

z	P(Z<z)	z	P(Z<z)	z	P(Z<z)	z	P(Z<z)	z	P(Z<z)
0.00	0.5000	0.50	0.6915	1.00	0.8413	1.50	0.9332	2.00	0.9772
0.01	0.5040	0.51	0.6950	1.01	0.8438	1.51	0.9345	2.02	0.9783
0.02	0.5080	0.52	0.6985	1.02	0.8461	1.52	0.9357	2.04	0.9793
0.03	0.5120	0.53	0.7019	1.03	0.8485	1.53	0.9370	2.06	0.9803
0.04	0.5160	0.54	0.7054	1.04	0.8508	1.54	0.9382	2.08	0.9812
0.05	0.5199	0.55	0.7088	1.05	0.8531	1.55	0.9394	2.10	0.9821
0.06	0.5239	0.56	0.7123	1.06	0.8554	1.56	0.9406	2.12	0.9830
0.07	0.5279	0.57	0.7157	1.07	0.8577	1.57	0.9418	2.14	0.9838
0.08	0.5319	0.58	0.7190	1.08	0.8599	1.58	0.9429	2.16	0.9846
0.09	0.5359	0.59	0.7224	1.09	0.8621	1.59	0.9441	2.18	0.9854
0.10	0.5398	0.60	0.7257	1.10	0.8643	1.60	0.9452	2.20	0.9861
0.11	0.5438	0.61	0.7291	1.11	0.8665	1.61	0.9463	2.22	0.9868
0.12	0.5478	0.62	0.7324	1.12	0.8686	1.62	0.9474	2.24	0.9875
0.13	0.5517	0.63	0.7357	1.13	0.8708	1.63	0.9484	2.26	0.9881
0.14	0.5557	0.64	0.7389	1.14	0.8729	1.64	0.9495	2.28	0.9887
0.15	0.5596	0.65	0.7422	1.15	0.8749	1.65	0.9505	2.30	0.9893
0.16	0.5636	0.66	0.7454	1.16	0.8770	1.66	0.9515	2.32	0.9898
0.17	0.5675	0.67	0.7486	1.17	0.8790	1.67	0.9525	2.34	0.9904
0.18	0.5714	0.68	0.7517	1.18	0.8810	1.68	0.9535	2.36	0.9909
0.19	0.5753	0.69	0.7549	1.19	0.8830	1.69	0.9545	2.38	0.9913
0.20	0.5793	0.70	0.7580	1.20	0.8849	1.70	0.9554	2.40	0.9918
0.21	0.5832	0.71	0.7611	1.21	0.8869	1.71	0.9564	2.42	0.9922
0.22	0.5871	0.72	0.7642	1.22	0.8888	1.72	0.9573	2.44	0.9927
0.23	0.5910	0.73	0.7673	1.23	0.8907	1.73	0.9582	2.46	0.9931
0.24	0.5948	0.74	0.7704	1.24	0.8925	1.74	0.9591	2.48	0.9934
0.25	0.5987	0.75	0.7734	1.25	0.8944	1.75	0.9599	2.50	0.9938
0.26	0.6026	0.76	0.7764	1.26	0.8962	1.76	0.9608	2.55	0.9946
0.27	0.6064	0.77	0.7794	1.27	0.8980	1.77	0.9616	2.60	0.9953
0.28	0.6103	0.78	0.7823	1.28	0.8997	1.78	0.9625	2.65	0.9960
0.29	0.6141	0.79	0.7852	1.29	0.9015	1.79	0.9633	2.70	0.9965
0.30	0.6179	0.80	0.7881	1.30	0.9032	1.80	0.9641	2.75	0.9970
0.31	0.6217	0.81	0.7910	1.31	0.9049	1.81	0.9649	2.80	0.9974
0.32	0.6255	0.82	0.7939	1.32	0.9066	1.82	0.9656	2.85	0.9978
0.33	0.6293	0.83	0.7967	1.33	0.9082	1.83	0.9664	2.90	0.9981
0.34	0.6331	0.84	0.7995	1.34	0.9099	1.84	0.9671	2.95	0.9984
0.35	0.6368	0.85	0.8023	1.35	0.9115	1.85	0.9678	3.00	0.9987
0.36	0.6406	0.86	0.8051	1.36	0.9131	1.86	0.9686	3.05	0.9989
0.37	0.6443	0.87	0.8078	1.37	0.9147	1.87	0.9693	3.10	0.9990
0.38	0.6480	0.88	0.8106	1.38	0.9162	1.88	0.9699	3.15	0.9992
0.39	0.6517	0.89	0.8133	1.39	0.9177	1.89	0.9706	3.20	0.9993
0.40	0.6554	0.90	0.8159	1.40	0.9192	1.90	0.9713	3.25	0.9994
0.41	0.6591	0.91	0.8186	1.41	0.9207	1.91	0.9719	3.30	0.9995
0.42	0.6628	0.92	0.8212	1.42	0.9222	1.92	0.9726	3.35	0.9996
0.43	0.6664	0.93	0.8238	1.43	0.9236	1.93	0.9732	3.40	0.9997
0.44	0.6700	0.94	0.8264	1.44	0.9251	1.94	0.9738	3.50	0.9998
0.45	0.6736	0.95	0.8289	1.45	0.9265	1.95	0.9744	3.60	0.9998
0.46	0.6772	0.96	0.8315	1.46	0.9279	1.96	0.9750	3.70	0.9999
0.47	0.6808	0.97	0.8340	1.47	0.9292	1.97	0.9756	3.80	0.9999
0.48	0.6844	0.98	0.8365	1.48	0.9306	1.98	0.9761	3.90	1.0000
0.49	0.6879	0.99	0.8389	1.49	0.9319	1.99	0.9767	4.00	1.0000
0.50	0.6915	1.00	0.8413	1.50	0.9332	2.00	0.9772		

Percentage points of the normal distribution

The values z in the table are those which a random variable $Z \sim N(0,1)$ exceeds with probability p; that is, $P(Z > z) = p$.

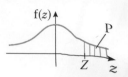

p	z	p	z
0.5000	0.0000	0.0500	1.6449
0.4000	0.2533	0.0250	1.9600
0.3000	0.5244	0.0100	2.3263
0.2000	0.8416	0.0050	2.5758
0.1500	1.0364	0.0010	3.0902
0.1000	1.2816	0.0005	3.2905

Answers

Exercise 1A

1 Using a mathematical model is quicker and cheaper. A mathematical model is a simplification which does not reflect all the aspects of the real problem.

2 Predictions based on the model are compared with observed data.
In the light of this comparison, the model is adjusted.
This process is repeated.

Exercise 2A

1 **a** quantitative **b** qualitative
 c quantitative **d** quantitative
 e qualitative

2 **a** discrete **b** continuous
 c discrete **d** continuous
 e continuous **f** continuous

3 **a** It is descriptive rather than numerical.
 b It is quantitative because it is numerical. It is discrete because its value must be an integer; you cannot have bits of a pupil.
 c It is quantitative because it is numerical. It is continuous because weight can take any value in a given range.

4 **a**

Height (cm)	Frequency	Cumulative frequency
165	8	8
166	7	15
167	9	24
168	14	38
169	18	56
170	16	72

 b 38 boys **c** 169 cm

5 **a**

Lifetime (hours)	Frequency	Cumulative frequency
5.0–5.9	5	5
6.0–6.9	8	13
7.0–7.9	10	23
8.0–8.9	22	45
9.0–9.9	10	55
10.0–10.9	2	57

 b 5.95 and 6.95 hours **c** 9.45 hours

6 **a** 1.4 kg and 1.5 kg **b** 1.35 kg

7 A is not true; B is true; C is true; D is true.

Exercise 2B

1 **a** 700 g **b** 600 g **c** 700 g
 d The mean will increase; the mode will remain unchanged; the median will decrease.

2 **a** 42.7
 b The mean will increase.

3 **a** 8 minutes **b** 10.2 minutes
 c 8.5 minutes
 d The median would be best. The mean is affected by the extreme value 26.

4 increase to 24.5 litres

5 5.7 days (1 d.p.)

6 **a** 4.3 mm **b** 26.5 hours
 c modal rainfall 3 mm, modal sunshine 15 hours
 d median rainfall 2.5 mm, median sunshine 16.5 hours
 e median rainfall and mean sunshine (least rainfall and highest sunshine)

7 71% (nearest percent)

Exercise 2C

1 **a**

Mark	5	6	7	8	9	10
Frequency	4	6	10	6	7	3

 b 7.42 **c** 16 **d** The mean is greater

2 5 eggs (mode 5, median 5, mean 5.44)

3 **a** 98 **b** 6 **c** 6.31
 d 6 **e** 6

4 **a** 36 **b** 31 **c** 2
 d 1 **e** 1.47 **f** the median

5 The company would use the mode (£48), since it is lower than the median (£54) and the mean (£56.10).

Exercise 2D

1 **a** 1.19 **b** 1.08
 c The hotel should consider getting a new lift since both the mean and the median are greater than 1.

2 **a** £351 to £400 **b** £345 **c** £355

3 **a** 82.3 decibels **b** 16

4 Store B (mean 51 years) employs older workers than store A (mean 50 years).

5 **a** 51 mph to 60 mph
 b 0.71 mph (2 d.p.) (mean 50.03 mph, median 50.74 mph)
 c 8% or 9%

Exercise 2E

1. 70
2. 48.5
3. a 3.5   b i 7   ii 35   iii 37
4. 365
5. a

Age (a)	Frequency (f)	Mid-point (x)	y
11–21	11	16	1.0
21–27	24	24	5.0
27–31	27	29	7.5
31–37	26	34	10.0
37–43	12	40	13.0

   b 29.0

Mixed Exercise 2F

1. a 50   b 50   c 54
2. 69.2
3. a mean £19.57, mode £6.10, median £7.80
   b The value £91.00 is wrong.
4. a The mean is higher than it should be.
   b 34.4
5. 607
6. £18 720
7. a group A 63.4, group B 60.2
   b The method used for group A may be better.
8. a 7.09 km   b 7.04 km
9. a 25.5 minutes
   b 26.6 minutes
   c She spent more time each week playing computer games in the last 40 days than in the first 50 days.
10. a 21 to 25 hours   b 21.6 hours
    c 20.6 hours   d 20.8 hours

Exercise 3A

1. a 7   b 9   c 4
2. a £290   b $Q_1 = 400$, $Q_3 = 505$.   c £105
3. a 25, 35, 55, 65, 90, 100; total 100
   b $Q_1 = 0.5$, $Q_3 = 4$
   c 3.5 hours
4. 1 ($Q_1 = 9$, $Q_3 = 10$)
5. a 3, 9, 19, 26, 31   b 389 kg
   c 480 kg   d 90.8 kg
6. a 1100   b 1833   c 733
7. a 71   b 24.6

Exercise 3B

1. a 8, 20, 56, 74, 89, 99
   b 10   c 8   d 9
2. a 11, 46, 80, 96, 106, 111
   b £17.10   c £28.25   d £11.15
3. £81.90
4. 6.2 minutes
5. a 49   b 38.7 minutes
   c 48.8 minutes

Exercise 3C

1. a 3   b 0.75   c 0.866
2. 3.11 kg

3. a 178 cm   b 59.9 cm^2 c 7.74 cm
4. mean 5.44, standard deviation 2.35
5. a 25   b 4
6. a The mean for both routes is 14.
   b Route 1 has variance 4 and standard deviation 2. Route 2 has variance 5.33 and standard deviation 2.31.
   c Route 1 would be best. Although the means are the same, the standard deviation for route 1 is lower, so this route is more reliable.

Exercise 3D

1. 133
2. 7.35
3. a

Number of £'s (x)	Number of students (f)	fx	fx^2
8	14	112	896
9	8	72	648
10	28	280	2800
11	15	165	1815
12	20	240	2880
Totals	85	869	9039

   b 1.82   c £1.35

4. a

Number of days absent (x)	Number of students (f)	fx	fx^2
0	12	0	0
1	20	20	20
2	10	20	40
3	7	21	63
4	5	20	80

   b 1.51   c 1.23

5. a

Lifetime in hours	Number of parts	Mid-point x	fx	fx^2
$5 < h \leq 10$	5	7.5	37.5	281.25
$10 < h \leq 15$	14	12.5	175.0	2187.50
$15 < h \leq 20$	23	17.5	402.5	7043.75
$20 < h \leq 25$	6	22.5	135.0	3037.50
$25 < h \leq 30$	2	27.5	55.0	1512.50

   b variance 22.0, standard deviation 4.69 hours
6. variance 21.25, standard deviation 4.61

Exercise 3E

1. a 5.08   b i 5.08   ii 5.08   iii 5.08
2. i 70.7   ii 70.7   iii 70.7
3. a 0.28   b 0.675   c 2.37   d 6.5
4. 2.34
5. 1.76 hours
6. 22.9
7. 416

Mixed Exercise 3F

1 **a** 6   **b** 3   **c** 9   **d** 6
2 37.5
3 **a** 20.5   **b** 34.7   **c** 14.2
4 15.5 m
5 **a** 40.9   **b** 54   **c** 13.1   **d** 10.1
6 **a** mean 15.8, standard deviation 2.06
   b The mean wing span will decrease.
7 **a** 98.75   **b** 104   **c** 5.58   **d** 4.47

Exercise 4A

1

										Key: 2|3 means 23 DVDs
0	6	9									(2)
1	2	2	2	5	5	5	7	8	9		(9)
2	0	2	3	5	5	5	6	6	7	7 9 9	(12)
3	2	2	4	4	5						(5)
4	2	5									(2)

   a 25   **b** 15   **c** 29
2 **a** 24   **b** 49   **c** 8   **d** 3
   e 37   **f** 34   **g** 21   **h** 37
3 **a** 41   **b** 32   **c** 47   **d** 15
   e 47
4 **a**

	Boys		Girls	
(2)	9 8	2	4 6 8	(3)
(3)	4 2 2	3	2 3 4 4 9	(5)
(5)	8 7 5 5 4	4	5 6 7	(3)
(5)	7 6 6 4 4	5	2 4	(2)
(1)	0	6		

   Key: 2|3|4 means 32 boys and 34 girls.
   b The girls gained lower marks than the boys.
5 **a** 17 males, 15 females
   b £48
   c Males earned more in general.

Exercise 4B

1 **a** 7 is an outlier.
   b 88 is not an outlier.
   c 105 is an outlier.
2 **a** no outliers
   b 170 g and 440 g
   c 760 g

Exercise 4C

1

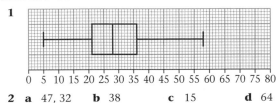

0	5	10	15	20	25	30	35	40	45	50	55	60	65	70	75	80

2 **a** 47, 32   **b** 38   **c** 15   **d** 64

Exercise 4D

1 **a** 45   **b** lower quartile
   c Boys have a lower median and a larger spread.
   or Girls have a higher median and a smaller spread.

2 **a** The male turtles have a higher median weight, a greater interquartile range and a greater total range.
   b It is more likely to have been female. Very few of the male turtles weighed this little, but more than a quarter of the female turtles weighed more than this.
   c 500 g

Exercise 4E

1 **a**

Height (cm)	Frequency	Class width	Frequency density
135–144	40	10	4
145–149	40	5	8
150–154	75	5	15
155–159	65	5	13
160–174	60	15	4

   b

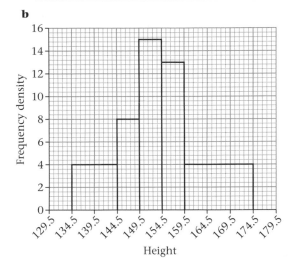

2 **a** The quantity (time) is continuous.
   b 150   **c** 369   **d** 699
3 **a** 114   **b** 90   **c** 24
4 **a** The quantity (distance) is continuous.
   b 310   **c** 75   **d** 95   **e** 65
5 **a** The quantity (weight) is continuous.
   b The area of the bar is proportional to the frequency.
   c 0.125   **d** 168   **e** 88

Exercise 4F

1 negative skew
2 **a** mean 31.1 minutes, variance 78.15
   b median 29.7 minutes; quartiles 25.8 minutes, 34.8 minutes
   c 0.475 (positive skew)
   d They will use the median because it is less than the mean.
3 **a** 64 mm
   b median 65 mm; quartiles 56 mm, 81 mm

c

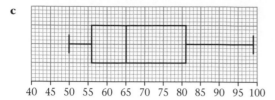

40 45 50 55 60 65 70 75 80 85 90 95 100

d & f The mean is greater than the median, so the data is positively skewed.

e mean 68.7 mm, standard deviation 13.7 mm

g various answers

Mixed Exercise 4G

1 a $Q_1 = 178$, $Q_2 = 185$, $Q_3 = 196$

  b 226

  c

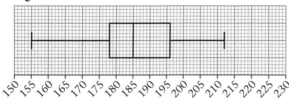

150 155 160 165 170 175 180 185 190 195 200 205 210 215 220 225 230

  d positive skew

2 a 22 **b** $X = 11$, $Y = 27$, $Z = 22$.

  c Strand Road has more pedal cycles, since its median is higher.

3 a It is true. 60 is the median for shop B.

  b It is true. 40 is the lower quartile for shop A.

  c Shop A has a greater interquartile range and a greater total range than shop B. Shop B has a higher median.

  d Shop B is more consistent.

4 a 45 minutes **b** 60 minutes

  c This represents an outlier.

  d Irt has a higher median than Esk. The interquartile ranges were about the same.

  e Esk positive skew, Irt symmetric

  f Esk had the fastest runners.

5 a 26 **b** 17

6 a 2.6 cm **b** 0.28 cm

7 a

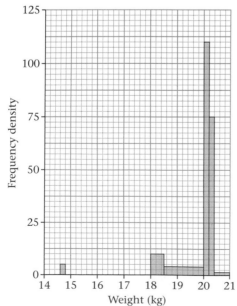

b mean 19.8 kg, standard deviation 0.963 kg

c 20.1 kg **d** −1.06 **e** negative skew

8 a 22.3

  b

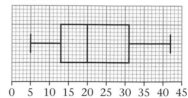

		Key: 1|3 means 13 bags				
0	5					(1)
1	0	1	3	5	7	(5)
2	0	0	5			(3)
3	0	1	3			(3)
4	0	2				(2)

  c median 20; quartiles 13, 31

  d no outliers

  e

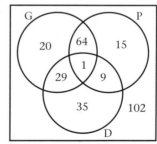

0 5 10 15 20 25 30 35 40 45

  f positive skew

Exercise 5A

1 0.5 **2** 0.5 **3** 0.25

4 0.125 **5** 0.0833 **6** 0.167

Exercise 5B

1 a 0.0769 **b** 0.25 **c** 0.0192

  d 0.308 **e** 0.75 **f** 0.231

2 a 0.56 **b** 0.24 **c** 0.32 **d** 0.04

3 a 0.6 **b** 0.1

4 0.4

5 a 0.12 **b** 0.08 **c** 0.08 **d** 0.432

6 a

  b 0.324 **c** 0.375 **d** 0.255 **e** 0.371

Exercise 5C

1 a 0.6 **b** 0.8 **c** 0.4 **d** 0.9

2 a 0.3 **b** 0.6 **c** 0.8 **d** 0.9

3 a 0.25 **b** 0.5 **c** 0.65 **d** 0.1

4 a 0.15 **b** 0.45 **c** 0.55 **d** 0.25

  e 0.3

5 0.1

6 a 0.17 **b** 0.18 **c** 0.55

7 a 0.3 **b** 0.3

Exercise 5D

1 0.0769

2 a 0.333 **b** 0.667

3 a 0.182 **b** 0.727

4 a 0.7 **b** 0.667 **c** 0.8 **d** 0.4

5 a 0.5 **b** 0.3 **c** 0.3

Frequency density

Weight (kg)

6 a 0.3   **b** 0.35   **c** 0.4
7 a 0.0833   **b** 0.15   **c** 0.233   **d** 0.357
   e 0.643   **f** 0.783

Exercise 5E

1 a 0.625   **b** $\frac{1}{2}$ **c** 0.167   **d** 0.555
2 a 0.163   **b** 0.507
3 0.36
4 a 0.25   **b** 0.333
5 If the contestant sticks, their probability of winning is $\frac{1}{3}$. If they switch, the probability of winning is $\frac{2}{3}$. So they should switch. (This answer assumes that the host knows where the sports car is.)

Exercise 5F

1 a

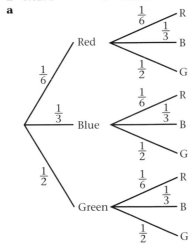

   b 0.7   **c** 0.3
2 a 0.05   **b** 0.2   **c** 0.6
3 a mutually exclusive
   b 0.6   **c** 0.4
4 a *various*   **b** *various*
5 a 0.391   **b** 0.625
6 a 0.0278   **b** 0.0217   **c** 0.290
7 a

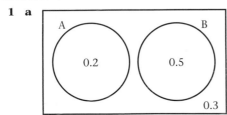
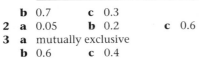

   b 0.389   **c** 0.611
8 a 0.0156   **b** 0.0911   **c** 0.0675   **d** 0.203

Mixed Exercise 5G

1 a *various*
   b

   c 0.333

2 a

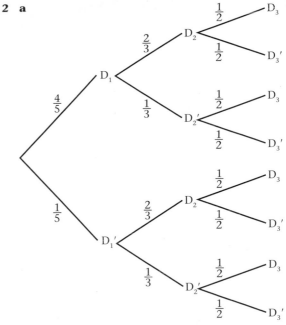

   b 0.267   **c** 0.233
3 a

   b 0.3   **c** 0.14   **d** 0.25
4 a

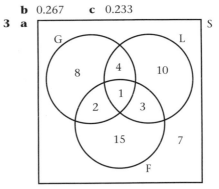

   b 0.3   **c** 0.25   **d** 0.2
5 a 0.123   **b** 0.231
6 a 0.5
   b

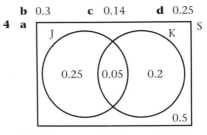

   c 0.895

7 a

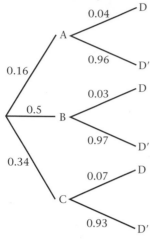

b 0.015   **c** 0.0452   **d** 0.332

8 a 0.2   **b** 0.5   **c** 0.245   **d** 0.571

9 a

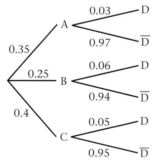

b 0.2   **c** 0.82   **d** 0.430   **e** 0.169

10 a 0.32   **b** 0.46   **c** 0.22   **d** 0.2016

Review Exercise 1

1 a

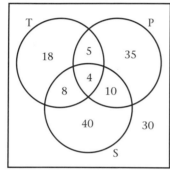

b i 0.0105   **ii** 0.0455
c 0.440

2 a positive skew
b median 26.7 miles
c mean 29.6 miles, standard deviation 16.6 miles
d 0.520
e yes; 0.520 > 0
f The median would be best, since the data is skewed.
g The distribution is symmetric (or has zero skew).

3 a Time is a continuous quantity.
b Area is proportional to frequency.
c *various*
d 30

4 a *any two of the following:*
Statistical models simplify a real world problem.
They are cheaper and quicker than an experiment.
They are easier to modify than an experiment.
They improve understanding of problems in the real world.
They enable us to predict outcomes in the real world.
b 3. The model is used to make predictions.
4. Experimental data is collected.
7. The model is refined.

5 a Distance is a continuous quantity.
b 0.8, 3.8, 5.3, 3.7, 0.75, 0.1
c $Q_2 = 58.8$, $Q_1 = 52.5$, $Q_3 = 67.1$.
d 62.5 km
e 0.137, positive skew
f The mean is greater than the median.

6 a

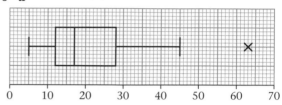

b The distribution is positively skewed, since $Q_2 - Q_1 < Q_3 - Q_2$
c Most of the delays are so small that passengers should find them acceptable.

7 a 0.338   **b** 0.46   **c** 0.743   **d** 0.218

8 a 56
b $Q_1 = 35$, $Q_2 = 52$, $Q_3 = 60$
c mean = 49.4, s.d. = 14.6
d -1.356
e The mean (49.4) is less than the median (52), which is less than the mode (56).

9 a

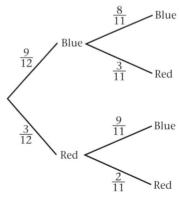

b 0.25   **c** 0.182

10 a

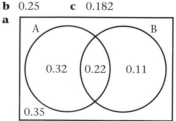

b $P(A) = 0.54$, $P(B) = 0.33$
c 0.478
d They are not independent.

11 a maximum, minimum, median, quartiles, outliers
  b i 37 minutes
    ii upper quartile, third quartile, 75 percentile
  c outliers – values that are much greater than or much less than the other values and need to be treated with caution
  d

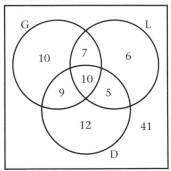

Time (minutes)

  e The children from school A generally took less time than those from school B.
The median for B is less than the median for A.
A has outliers, but B does not.
Both have positive skew.
The interquartile range for A is less than the interquartile range for B.
The total range for A is greater than the total range for B.

12 a 0.0370
  b 19.3 minutes
  c 24.8 minutes
  d Their conversations took much longer during the final 25 weeks.

13 a

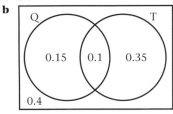

  b 0.1   **c** 0.41   **d** 0.21   **e** 0.667
14 a 0.1
  b

  c 0.25
15 a 35, 15
  b 40
  c 18.9 minutes
  d 7.26 minutes
  e median 18 minutes; quartiles 13.75 minutes, 23 minutes
  f 0.376, positive skew

16 a

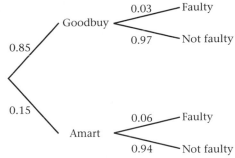

  b 0.9655
17 mean 240, standard deviation 14

Exercise 6A

1 a i no correlation
    ii negative correlation
    iii positive correlation
  b i There is no correlation between height and intelligence.
    ii As age increases, price decreases.
    iii As length increases, breadth increases.
2 a positive correlation
  b The longer the treatment, the greater the loss of weight.
3 a

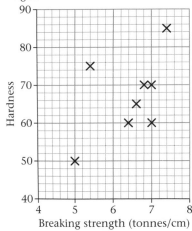

  b There is weak positive correlation. There is some reason to believe that, as breaking strength increases, hardness increases.
4 a

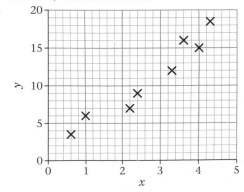

positive correlation

b

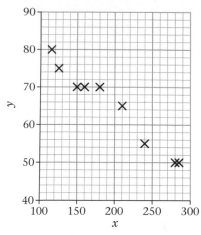

negative correlation

5 a

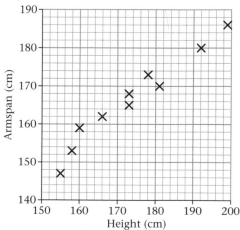

b There is positive correlation. As height increases, arm-span increases.

6 a

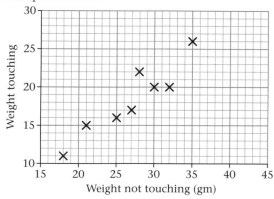

b There is positive correlation. If a student guessed a greater weight before touching the bag, they were more likely to guess a greater weight after touching it.

Exercise 6B

1 1.775
2 7.90
3 −14
4 0.985
5 0.202

6 a 9.71
  b 0.968
  c There is positive correlation. The greater the age, the taller the person.
7 a $S_{ll} = 30.3$, $S_{tt} = 25.1$, $S_{lt} = 25.35$
  b 0.919
8 a 0.866
  b There is positive correlation. The higher the IQ, the higher the mark in the intelligence test.
9 a $S_{xx} = 82.5$, $S_{yy} = 32.9$, $S_{xy} = -44.5$
  b −0.854
  c There is negative correlation. The relatively older young people took less time to reach the required level.

Exercise 6C

1 a iii (0)   **b i** (−0.96)
2 a i (−1)   **b iii** (0)
3 There is strong positive correlation. The taller the father is, the taller his son will be.
4 a ii   **b iv**   **c iii**   **d i**
5 There is strong positive correlation between x and y. As x increases, y increases.
There is strong negative correlation between s and t. As s increases, t decreases.
6 This is not sensible. There is no way in which one could be directly caused by the other.
7 This is not sensible. It is more likely that the taller pupils are older.

Exercise 6D

1 a $x - 2000$, $\dfrac{y}{3}$
  b $\dfrac{s}{100}$, no change to t
2 0.973
3 0.974
4 a $S_{pp} = 10$, $S_{tt} = 5.2$, $S_{pt} = 7$
  b 0.971
  c 0.971
5 a

x	15	37	5	0	45	27	20
y	30	13	34	43	20	14	0

  b $S_{xx} = 1.59$, $S_{yy} = 62$, $S_{xy} = 8.1$
  c 0.816
  d 0.816
  e The greater the mass of a woodmouse, the longer its tail.
6 a $S_{xx} = 1601$, $S_{yy} = 1282$, $S_{xy} = -899$
  b −0.627
  c The shopkeeper is wrong. There is negative correlation. Sweet sales actually decrease as newspaper sales increase.

Mixed Exercise 6E

1 a

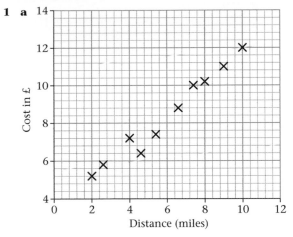

b There is correlation positive. The further the taxi travels, the more it costs.

2 a i shows positive correlation.
   ii shows negative correlation.
   iii shows no correlation.
  b i The older a snake is, the longer it is.
   ii The higher the unemployment, the lower the drop in wages.
   iii There is no correlation between the age and height of men.

3 a ii (-0.12)
  b i (0.87)
  c iii (-0.81)

4 As a person's age increases, their score on the memory test decreases.

5 a -0.147 **b** -0.147
  c This is weak negative correlation. There is just a little evidence to suggest that students in the group who are good at science are also good at art.

6 a $S_{jj} = 4413$, $S_{pp} = 5145$, $S_{jp} = 3972$
  b 0.834
  c There is strong positive correlation, so Nimer is correct.

7 a

x	20	40	30	52	15	5	80	40	5	0
y	7	10	9	10.5	6	5	10	9	6	3

  b $S_{xx} = 5642$, $S_{yy} = 57.2$, $S_{xy} = 494$
  c 0.870
  d 0.870
  e This is a strong positive correlation. As v increases, m increases.

8 a

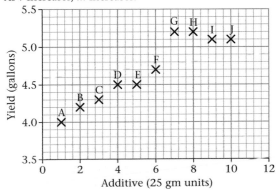

b There is some positive correlation. Using the additive increases milk yield.
c Each cow should be given 7 units. The yield levels off at this point.
d The yield stops rising after the 7th cow (G).
e 0.952
f It would be less than 0.952. The yield of the last 3 cows is no greater than that of the 7th cow.

9 a

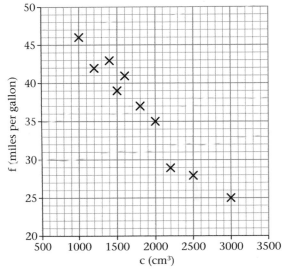

b There is negative correlation. As engine size increases, the number of miles per gallon decreases.
c $S_{cf} = -38\,200$
d $\dfrac{c}{100}$ or $\dfrac{(c-1000)}{100}$ and $f - 25$.

10 a $S_{xx} = 91.5$, $S_{yy} = 38.9$, $S_{xy} = 32.3$
  b 0.541
  c 0.541
  d There is positive correlation. As age increases, blood pressure increases.

Exercise 7A

1 The number of operating theatres is the independent variable.
The number of operations is the dependent variable.
2 The number of suitable habitats is the independent variable.
The number of species is the dependent variable.
3 $a = -3$, $b = 6$.
4 $y = -14 + 5.5x$
5 $y = 2x$
6 a $S_{xx} = 5$, $S_{xy} = 20$
  b $y = 2 + 4x$
7 a $S_{xx} = 40.8$, $S_{xy} = 69.6$
  b $y = -0.294 + 1.71x$
8 $g = 1.50 + 1.44h$
9 a $S_{nn} = 6486$, $S_{np} = 6344$
  b $p = 20.9 + 0.978n$
10 a $S_{xx} = 10$, $S_{xy} = 14.5$
  b $y = -0.07 + 1.45x$

Exercise 7B

1 $y = 6 - x$
2 $s = 88 + p$

3 $y = 32 - 5.33x$

4 $t = 9 + 3s$

5 **a** $y = 3.5 + 0.5x$
  b $d = 35 + 2.5c$

6 **a** $S_{xy} = 162, S_{xx} = 191; y = 7.87 + 0.85x$
  b $y = 22.35 + 2.125h$

Exercise 7C

1 6

2 384 g

3 **a** Extrapolation means using the regression line to estimate outside the range of the data. It can be unreliable.
  b Interpolation means using the regression line to estimate within the range of the data. It is usually reasonably reliable.

4 **a** £6.00. This is reliable since 7 years is within the range of the data used.
  b 3 years is outside the range of the data.

5 This is not a sensible estimate since 30 minutes is a long way outside the range of the data.

6 **a** 3845. This is reliable since £2650 is within the range of the data.
  b 9730. This is unreliable since £8000 is outside the range of the data.
  c 1100 extra books are sold for each £1000 spent on advertising.
  d It suggests that 930 books would be sold if no money were spent on advertising. This is reasonably reliable since it is only just outside the range of the data.

7 **a** 283
  b If dexterity increases by 1 unit, production increases by 57 units.
  c **i** The estimate would be reliable. 2 is outside the range of the data, but only just.
    ii The estimate would be unreliable. 14 is a long way outside the range of the data.

8 The equation of the regression line would change.

Mixed Exercise 7D

1 **a** 2.45 mm
  b 4.52 mm
  c The answer to part a is reliable since 300°C is within the range of the data. The answer to part **b** is unreliable since 530°C is outside the range of the data.

2 **a** $t = 1.96 + 0.95s$
  b 49.46

3 **a** $S_{xx} = 6.43, S_{xy} = 11.7$
  b $y = 0.554 + 1.82x$
  c 6.01 cm

4 **a** $S_{xx} = 16\,350, S_{xx} = 210\,330$
  b $y = 225 + 12.9x$
  c 1510
  d 255
  e This answer is unreliable since a gross national product of 3500 is long way outside the range of the data.

5 **a** $y = 0.343 + 0.499x$
  b $t = 2.34 + 0.224m$
  c 4.58 cm

6 **a** $S_{xx} = 90.9, S_{xx} = 190$
  b $y = -1.82 + 2.09x$

c $p = 25.5 + 2.09r$
d 71.4
e This answer is reliable since 22 breaths per minute is within the range of the data.

7 **a** 0.79 kg is the average amount of food consumed in 1 week by 1 hen.
  b 23.9 kg
  c £47.59

8 **a & e**

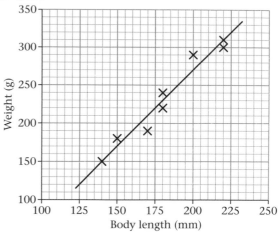

  b There appears to be a linear relationship between body length and body weight.
  c $w = -12.7 + 1.98l$
  d $y = -127 + 1.98x$
  f 289 g. This is reliable since 210 mm is within the range of the data.
  g Water voles B and C were probably removed from the river since they are both underweight. Water vole A was probably left in the river since it is slightly overweight.

9 **a** $S_{xy} = 78, S_{xx} = 148$
  b $y = 7.31 + 0.527x$
  c $w = 816 + 211n$
  d 5036 kg
  e 100 items is a long way outside the range of the data.

Exercise 8A

1 **a** This is not a discrete random variable, since time is a continuous quantity.
  b This is not a discrete random variable, since it is always 7 and thus does not vary
  c This is a discrete random variable, since it is always a whole number and it does vary.

2 0, 1, 2, 3, 4

3 **a** {(2, 2), (2, 3), (3, 2), (3, 3)}
  b

x	4	5	6
$P(X = x)$	0.25	0.5	0.25

  c $P(X = 4) = 0.25$, $P(X = 5) = 0.5$, $P(x = 6) = 0.25$.

4 0.0833

5 $k + 2k + 3k + 4k = 1$,
  so $10k = 1$, so $k = 0.1$.

6

x	1	2	3	4	5
$P(X = x)$	0	0.1	0.2	0.3	0.4

7 a 0.125

b

x	1	2	3	4
$P(X = x)$	0.125	0.125	0.325	0.325

8 a 0.3

b

x	-2	-1	0	1	2
$P(X = x)$	0.1	0.1	0.3	0.3	0.2

9 0.25

Exercise 8B

1 a 0.5 **b** 0.2 **c** 0.6

2 a 0.625 **b** 0.375

3 a

x	1	2	3	4	5	6
$F(x)$	0.1	0.2	0.35	0.60	0.9	1.0

b 0.9 **c** 0.2

4 a

x	1	2	3	4	5	6
$P(X = x)$	0.1	0.1	0.25	0.05	0.4	0.1

b 0.5 **c** 0.4

5 a 0.0556

b

x	$P(X = x)$
1	0.0556
2	0.0556
3	0.1667
4	0.1667
5	0.2777
6	0.2777

c 0.389 **d** 0.444 **e** 0.0556

6 a 0.25

b

x	-2	-1	0	1	2
$P(X = x)$	0.1	0.1	0.25	0.25	0.3

c 0.45

7 a 0.833 **b** 0.167

c

x	1	2	3	4	5
$P(X = x)$	0.333	0.167	0.167	0.167	0.167

8 a 1

b

x	1	2	3
$P(X = x)$	0.25	0.3125	0.4375

Exercise 8C

1 a $E(X) = 4.6$, $E(X^2) = 26$

b $E(X) = 2.8$, $E(X^2) = 9$.

2 $E(X) = 4$, $E(X^2) = 18.2$.

3 a

x	2	3	6
$P(X = x)$	0.5	0.333	0.167

b $E(X) = 3$, $E(X^2) = 11$

c no

4 a

Number of heads (h)	0	1	2
$P(H = h)$	0.25	0.5	0.25

b 0 heads 12.5 times, 1 head 25 times, 2 heads 12.5 times

c The coins may be biased. There were more times with 2 heads and fewer times with 0 heads than expected.

5 a = 0.3, b = 0.3.

6 100

Answers 8D

1 a 1 **b** 2

2 a $E(X) = 1.83$, $Var(X) = 0.472$

b $E(X) = 0$, $Var(X) = 0.5$

c $E(X) = -0.5$, $Var(X) = 2.25$.

3 $E(Y) = 3.5$, $Var(X) = 0.917$.

4 a

x	$P(X = x)$
2	0.0278
3	0.0556
4	0.0833
5	0.1111
6	0.1389
7	0.1667
8	0.1389
9	0.1111
10	0.0833
11	0.0556
12	0.0278

b 7

c 5.833

5 a

d	0	1	2	3
$P(D = d)$	0.25	0.375	0.25	0.125

$P(D = 3) = 0.125$.

b 1.25

c 0.9375

6 a $P(T = 1) = P(\text{head}) = 0.5$,
$P(T = 2) = P(\text{tail, head}) = 0.5 \times 0.5 = 0.25$,
$P(T = 3) = 1 - P(T = 1) - P(T = 2) = 0.25$.

b $E(T) = 1.75$, $Var(T) = 0.687$.

7 a 2

b $a = 0.375$, $b = 0.25$.

Answers 8E

1 a 8 **b** 40

2 a 6 **b** 7 **c** 1 **d** 0

e 54 **f** 54 **g** 6

3 a 7 **b** 5 **c** 36 **d** 9

4 a 4μ **b** $2\mu + 2$ **c** $2\mu - 2$ **d** $4\sigma^2$

e $4\sigma^2$

5 a 7 **b** -4 **c** 81 **d** 81

e 13 **f** 12

6 $E(S) = 64$, $Var(S) = 225$.

7 a

x	1	2	3
P($X = x$)	0.25	0.375	0.375

  b 2.125   **c** 0.609   **d** 5.25   **e** 5.48

8 a 0.2   **b** 0.76   **c** 1.07   **d** 0.0844

Exercise 8F

1 E(X) = 3, Var(X) = 2.

2 a 4   **b** 4

3 a E(X) = 3.5, Var(X) = 2.92
  b 0.667

4 a 0.3   **b** E(X) = 11, Var(X) = 33

5 a 0.2   **b** E(X) = 10, Var(X) = 33

6 A discrete uniform distribution is not a good model. The game depends on the skill of the player. The points are likely to cluster around the middle.

7 a a discrete uniform distribution
  b 4.5
  c 5.25
  d The expected winnings are less than the 5p stake.

Mixed Exercise 8G

1 a

x	P($X = x$)
1	0.0476
2	0.0952
3	0.1429
4	0.1905
5	0.2381
6	0.2857

  b 0.571   **c** 4.33   **d** 2.22   **e** 8.89

2 a 0.2   **b** 0.7   **c** 0.6   **d** 3.6
  e 8.04

3 a 0.3
  b E(X) = 0 × 0.2 + 1 × 0.3 + 2 × 0.5 = 1.3.
  c 0.61
  d 0.5

4 a $k + 0 + k + 2k = 1$,
    so 4k = 1, so $k = 0.25$.
  b E(X) = 2
    E(X^2) = 0^2 × 0.25 + 1^2 × 0 + 2^2 × 0.25 + 3^2 × 0.5
    = 1 + 4.5 = 5.5.
  c 6

5 a 0.125   **b** 0.75   **c** 1.125   **d** 2.375
  e 0.859

6 a discrete uniform distribution
  b *any discrete distribution where all the probabilities are the same*
  c 2
  d 2

7 a $p + q = 0.5, 2p + 3q = 1.3$
  b $p = 0.2, q = 0.3$.
  c 1.29
  d 5.16

8 a 0.111
  b 3.44
  c Var(X) = E(X^2) − E(X)^2
    ~ 13.88889 − 3.44444^2
    ~ 2.02.
  d 8.1 (2 s.f.)

9 a E(X) = 3.5
    Var(X) = E(X^2) − E(X)^2
    = $\frac{91}{6} - \frac{49}{4} = \frac{35}{12}$
  b 6
  c 11.7

10 a

x	1	2	3	
P($X = x$)	0.0769	0.1923	0.3077	0.4231

  b 0.731
  c 3.077
  d Var(X) = E(X^2) − E(X)^2
    ~ 10.385 − 1.077^2 ~ 0.92
  e 8.28

Exercise 9A

1 a 0.9830 **b** 0.9131 **c** 0.2005 **d** 0.3520

2 a 0.1056 **b** 0.9535 **c** 0.0643 **d** 0.9992

3 a 0.9875 **b** 0.4222 **c** 0.4893 **d** 0.0516

4 a 0.0902 **b** 0.9438 **c** 0.1823 **d** 0.8836

Exercise 9B

1 a 1.33 **b** 1.86 **c** −0.42 **d** −0.49

2 a 2.50 **b** 0.22 **c** −0.71 **d** 0.8416

3 a 1.0364 **b** −1.6449 **c** 1.22 **d** 3.0902

4 a 1.06 **b** 2.55 **c** 0.81 **d** 1.35

5 a 0.2533 **b** 1.0364 **c** 1.2816 **d** 0.5244

Exercise 9C

1 a 0.9332   **b** 0.9772

2 a 0.0475   **b** 0.2514

3 a 0.1587   **b** 0.4985

4 a 0.264   **b** 0.171

5 a 0.961 *or* 0.962 **b** 22.4

6 32.6

7 18.1

8 a 19.1   **b** 18.3   **c** 0.0915

9 a 70.6   **b** 80.8   **c** 0.075

10 a 81.0 **b** 80.6   **c** 0.0364

Exercise 9D

1 11.5

2 3.87

3 31.6

4 25

5 $\mu = 13.1, \sigma = 4.32$.

6 $\mu = 28.3, \sigma = 2.59$.

7 $\mu = 12, \sigma = 3.56$.

8 $\mu = 35, \sigma = 14.8$ *or* $\sigma = 14.9$.

9 4.75

10 $\sigma = 1.99, a = 2.18$.

Mixed Exercise 9E

1 a 0.0401 **b** 188 cm

2 a 12.7% or 12.8%   **b** 51.1% or 51.2%

3 a 0.0668 **b** 0.052

4 a 3.65   **b** 0.1357   **c** 32.5

5 a 8.60 ml **b** 0.123   **c** 109 ml

6 $\mu = 30, \sigma = 14.8$ *or* $\sigma = 14.9$

7 mean 3.76 cm, standard deviation 10.2 cm

8 a 0.3085
  b 0.370 *or* 0.371
  c The first score was better, since fewer of the students got this score or more.
9 a 4.25 *or* 4.26   **b** 0.050
10 a 8.54 minutes   **b** 0.176
11 mean 6.12 mm, standard deviation 0.398 mm
12 a 0.8413   **b** 0.111
13 a 0.2119   **b** 28.2

Review Exercise 2

1 a £17
  b $S_{tm} = 1191.8$, $S_{tt} = 983.6$, $S_{mm} = 1728.9$
  c 0.914
  d 0.914. Linear coding does not affect the correlation coefficient.
  e 0.914 suggests a relationship between the time spent shopping and the money spent.
    0.178 suggests that there was no such relationship.
  f various

2 a

x	$P(X = x)$
1	0.0278
2	0.0833
3	0.1389
4	0.1944
5	0.2500
6	0.3056

  b 0.583   **c** 4.47   **d** 22.0   **e** 17.7
3 a 0.2743   **b** 12
4 Diagram A corresponds to −0.79, since there is negative correlation.
  Diagram B corresponds to 0.08, since there is no significant correlation.
  Diagram C corresponds to 0.68, since there is positive correlation.
5 a $y = -0.425 + 0.395x$
  b $f = 0.735 + 0.395m$
  c 93.6 litres
6 a 0.0588   **b** 3.76   **c** 1.47   **d** 13.3
7 a 0.076 *or* 0.077
  b 0.639 *or* 0.640
  c 153.2
8 a $p + q = 0.4$, $2p + 4q = 1.3$
  b $p = 0.15$, $q = 0.25$
  c 1.75
  d 7.00
9 a

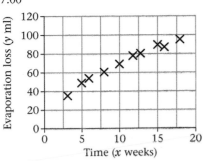

b The points lie close to a straight line.
c $a = 29.02$, $b = 3.90$.
d 3.90 ml of the chemicals evaporate each week.
e i 103 ml   **ii** 166 ml
f i This estimate is reasonably reliable, since it is just outside the range of the data.
  ii This estimate is unreliable, since it is far outside the range of the data.
10 a A statistical model simplifies a real world problem.
  It improves the understanding of a real world problem.
  It is quicker and cheaper than an experiment or a survey.
  It can be used predict possible future outcomes.
  It is easy to refine a statistical model.
  b i *various*   **ii** *various*
11 a 0.0618   **b** 0.9545
  c 0.00281   **d** This is a bad assumption.
12 a $S_{xy} = 71.47$, $S_{xx} = 1760$
  b $y = 0.324 + 0.0406x$
  c 2461.95 mm
  d $l = 2460.324 + 0.0406t$
  e 2463.98 mm
  f This estimate is unreliable since it is outside the range of the data.
13 a $E(X) = 3$
  $$Var(X) = \frac{(5 + 1)(5 - 1)}{12} = 2$$
  b 7
  c 18
14 a

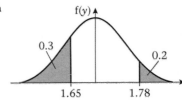

b mean 1.700 m, standard deviation 0.095 m
c 0.337
15 a −0.816
  b Houses are cheaper further away from the town centre.
  c −0.816
16 a & d

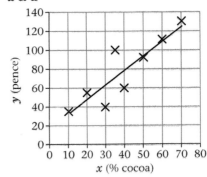

b $S_{xy} = 28\,750 - \dfrac{(315 \times 620)}{8} = 4337.5$, $S_{xx} = 2822$
c $a = 17.0$, $b = 1.54$
e i Brand D is overpriced, since it is a long way above the line.
  ii 69 *or* 70 pence

17 a 0.1056
  b 27.2
18 a $p + q = 0.45$, $3p + 7q = 1.95$
  b $p = 0.3$, $q = 0.15$
  c 0.35
  d 7.15
  e 1
  f 114.4
19 a 0.5
  b 54 kg
  c It represents an outlier or extreme value.
    It could be drums or a double base.
  d No skew
  e 13.3 kg *or* 13.4 kg

Examination style paper

1

3	<u>4</u>	<u>4</u>	<u>4</u>	<u>5</u>	<u>5</u>	<u>6</u>
2	<u>3</u>	<u>3</u>	<u>3</u>	<u>4</u>	<u>4</u>	<u>5</u>
2	<u>3</u>	<u>3</u>	<u>3</u>	<u>4</u>	<u>4</u>	<u>5</u>
1	2	2	2	<u>3</u>	<u>3</u>	<u>4</u>
1	2	2	2	<u>3</u>	<u>3</u>	<u>4</u>
1	2	2	2	<u>3</u>	<u>3</u>	<u>4</u>
Second First	1	1	1	2	2	3

P(Sum at least 3) $= \frac{27}{36} = \frac{3}{4}$

Alternative Solution
Let D_1 = the number on the first die
Let D_2 = the number on the second die
$$P(D_1 + D_2 \geqslant 3) = 1 - P(D_1 + D_2 = 2)$$
$$= 1 - P(D_1 = 1 \text{ and } D_2 = 1)$$
$$= 1 - P(D_1 = 1) \times P(D_2 = 1)$$
$$= 1 - \frac{1}{2} \times \frac{1}{2}$$
$$= \frac{3}{4}$$

2 a $P(X < 450) = P\left(Z < \dfrac{450 - 460}{10}\right) = P(Z < -1.0)$

  $= 1 - 0.8413 = 0.1587$
  b Expected number of jars $= 30 \times 0.1587$
    $= 4.761$ or 4.76 or 4.8
  c $P(X < 450) = 0.01$
  $\dfrac{450 - \mu}{10} = -2.3263$

  $\mu = 473.263 = 473$ to 3 s.f.
3 a $0.5 + b + 2a = 1$
    $0.3 + 2b + 6a = 1.6$
  Solving
    $a = 0.15$, $b = 0.2$
  b $E(5 - 2X) = 5 - 2E(X)$
    $= 5 - 2 \times 1.6 = 1.8$
  c $Var(X) = 1^2 \times 0.3 + 2^2 \times 0.2 + 3^2 \times 0.3 - 1.6^2$
    $= 1.24$

4 a mean $= \dfrac{302}{16} = 18.875$

  standard deviation is $\sqrt{\dfrac{5722}{16} - 18.875^2}$

  $= \sqrt{1.359375} = 1.16592\ldots$

  b mean % attendance is $\dfrac{18.875}{20} \times 100 = 94.375$

c Mode is 17
  Median is 18
  IQR is $20 - 17 = 3$
d First Group:

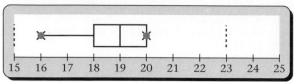

Second Group:

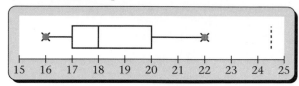

e First mean % > Second mean %
  First IQR < Second IQR
  First sd < Second sd
  First range < second range
  First negative skew, given by whiskers, symmetric by box
  Second positive skew.
5 a Discrete uniform distribution
  b $\dfrac{(n + 1)}{2} = 10$

    $n = 19$
  c $\dfrac{(n + 1)(n - 1)}{2} = 180$

6 a & e

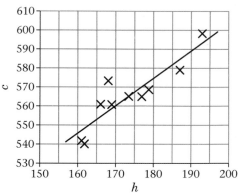

  b $S_{hh} = 272094 - \dfrac{1562^2}{9} = 1000.\dot{2}$

  $S_{cc} = 2878966 - \dfrac{5088^2}{9} = 2550$

  $S_{hc} = 884484 - \dfrac{1562 \times 5088}{9} = 1433.\dot{3}$

  c $r = \dfrac{S_{hc}}{\sqrt{S_{hh}S_{cc}}} = \dfrac{1433.\dot{3}}{\sqrt{1000.\dot{2} \times 2550}} = 0.897488$

  d $b = \dfrac{1433.\dot{3}}{1000.\dot{2}} = 1.433015$

    $a = \dfrac{5088}{9} - b \times \dfrac{1562}{9} = 316.6256$

    $c = 1.43h + 317$

  e See Graph
  f For every 1 cm increase in height, the confidence measure increases by 1.43.

g $h = 1.72$

$c = 1.43 \times 172 + 317 = 563$ to 3 s.f.

7 a P(Scores 15 points)

$= $ P(hit,hit,hit) $= 0.4 \times 0.4 \times 0.4 = 0.064$

b

x	0	5	10	15
P($X = x$)	0.6	0.4×0.6	$0.4^2 \times 0.6$	
	0.6	0.24	0.096	0.064

c P(Jean scores more in round two than round one)

$= $ P($X = 0$ then $X = 5$, 10 or 15)

$+ $ P($X = 5$ then $X = 10$ or 15)

$= $ P($X = 10$ then $X = 15$)

$= 0.6 \times (0.24 + 0.096 + 0.064)$

$+ 0.24 \times (0.096 + 0.064)$

$+ 0.096 \times 0.064$

$= 0.284544$

$= 0.285$ (3 s.f.)

Index